Second Edition

PT CLASSROOM TRAINING BOOK

The American Society for Nondestructive Testing

Copyright © 2019 by The American Society for Nondestructive Testing Inc.

The American Society for Nondestructive Testing, Inc. (ASNT) is not responsible for the authenticity or accuracy of information herein. Published opinions and statements do not necessarily reflect the opinion of ASNT. Products or services that are advertised or mentioned do not carry the endorsement or recommendation of ASNT.

No part of this publication may be reproduced or transmitted in any form, by means electronic or mechanical including photocopying, recording or otherwise, without the expressed prior written permission of The American Society for Nondestructive Testing, Inc.

IRRSP, *NDT Handbook*, *The NDT Technician*, and www.asnt.org are trademarks of The American Society for Nondestructive Testing, Inc. ACCP, ASNT, *Level III Study Guide*, *Materials Evaluation*, *Nondestructive Testing Handbook*, *Research in Nondestructive Evaluation,* and *RNDE* are registered trademarks of The American Society for Nondestructive Testing Inc.

Second Edition
   first printing 04/19
   ebook 04/19

Errata, if available for this printing, may be obtained from ASNT's website, www.asnt.org.

ISBN: 978-1-57117-445-1 (print)
ISBN: 978-1-57117-446-8 (ebook)

Printed in the United States of America

Published by:
The American Society for Nondestructive Testing Inc.
1711 Arlingate Lane
Columbus, OH 43228-0518
www.asnt.org

Publications Team:

Tim Jones, Senior Manager Publications

Editorial:
Toni Kervina, Educational Materials Editor
Cynthia M. Leeman, Educational Materials Supervisor

Production:
Joy Grimm, Production Manager
Synthia Jester, Graphic Designer and Illustrator

ASNT Mission Statement:
ASNT exists to create a safer world by advancing scientific, engineering, and technical knowledge in the field of nondestructive testing.

Acknowledgments

The American Society for Nondestructive Testing Inc. is grateful for the technical expertise, knowledge, and contributions of Brenda Collins and the team of liquid penetrant testing method reviewers, contributors, and subject matter experts who tirelessly worked to develop the second edition of the *Liquid Penetrant Testing Classroom Training Book*. The second edition builds on the first edition, originally written by Duane Badger in 2005.

Brenda Collins, Technical Editor – Sherwin Inc.
Adam Barrett – Naval Air Systems Command
Cody Bennett – US Steel Tubular Products Inc.
Charles Eick – Royal Blue NDT Services
Eric Henry – Applied Technical Services
Luis Payano – Port Authority of NY & NJ
Peter Pelayo – Met-L-Chek
Juan Carlos Ruiz-Rico – DNV GL
James Sieger – Valence Surface Technologies
Rusty Waldrop – US Coast Guard

The Publications Review Committee includes:

Joseph L. Mackin, Chair – International Pipe Inspectors Association
Martin T. Anderson – Alaska Technical Training
Mark R. Pompe – West Penn Testing Group

Foreword

Purpose

The American Society for Nondestructive Testing, Inc. (ASNT) has prepared this series of Personnel Training Publications to provide an overview in a classroom setting of a given nondestructive testing (NDT) method. Each classroom training book in the series is organized to follow the body of knowledge found in *ANSI/ASNT CP-105: ASNT Standard Topical Outlines for Qualification of Nondestructive Testing Personnel* (2016). Level I and Level II candidates should use this classroom training book as a preparation tool for NDT certification. Note, however, that an NDT Level I or Level II may be expected to know additional information based on industry or employer requirements.

Supplementary Material

Although the classroom training book may be purchased and read as a standalone product, it is intended to be used in conjunction with the Lecture Guide and PowerPoint presentation for instructors and Student Guide for students. These guides contain a condensed version of the material in the classroom training book and quiz questions per chapter (lesson) for review purposes.

Contents

Acknowledgments iii

Foreword iv

LEVEL I 1

Chapter 1: Introduction to Liquid Penetrant Testing 3
History 3
The Purpose of Liquid Penetrant Testing 4
Basic Principles 4
Safety Precautions 8

Chapter 2: Liquid Penetrant Processing 11
Procedures and Techniques 11
Precleaning and Post-cleaning Equipment 12
Application of Penetrant 15
Removal of Excess Surface Penetrant 15
Interpretation and Evaluation 20
Indications 21
Categories of Relevant Indications 22
Discontinuity Depth Determination 24

Chapter 3: Liquid Penetrant Testing Methods 25
Standard Methods 25

Chapter 4: Liquid Penetrant Testing Equipment 33
Liquid Penetrant Testing Units 33
Portable Equipment 38
Ultraviolet Radiation 39
Materials For Liquid Penetrant Testing 43
Additional Precautions 46

LEVEL II 47

Chapter 5: Selection of Liquid Penetrant Testing Method 49
Selection 49
Advantages and Disadvantages 50

Chapter 6: Interpretation and Evaluation of Indications 55
Discontinuity Categories 55
Indications 58
Factors Affecting Indications 60
Crack Indications 62
Porosity Indications 63

Indications from Specific Material Forms 63
Evaluation of Indications 69

Chapter 7: Liquid Penetrant Process Control 73
Quality Control of Test Materials 73
Reference Blocks 73
Liquid Penetrant Material Tests 78
Emulsifier Tests 80
Developer Tests 80

Chapter 8: Test Procedures and Standards 81
Procedures, Standards, and Codes 81
Basic Methods of Instruction 82
Conclusion 83

References 87

Figure Sources 89

Glossary 91

Index 95

LEVEL >>

Introduction to Liquid Penetrant Testing

Liquid penetrant testing (PT) is a versatile nondestructive testing (NDT) method used for detecting discontinuities open to the surface in a wide variety of solid, nonporous materials. The effectiveness of the test is determined by the training, skill, and dedication of the penetrant technician; the cleaning and preparation of the test object; and the materials and procedures used to perform the test.

History

The oil-and-whiting method used in the railroad industry in the early 1900s was the first recognized use of the principles of penetrants to detect cracks (Figure 1). The method used an oil solvent for cleaning followed by the application of a whiting or chalk coating, which absorbed oil from the cracks to reveal their locations. A dye was then added to the liquid. By the 1940s, fluorescent or visible dye was added to the oil used to penetrate test objects.

Experience showed that temperature and soak time were important. This started the practice of written instructions to provide standard, uniform results. The use of written procedures has evolved, giving the ability for design engineers and manufacturers to get the same high standard results from any properly trained and certified PT technician.

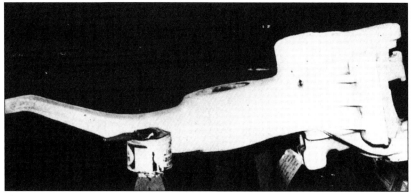

Figure 1. In the early 1930s, the oil-and-whiting method being used in a railroad shop on a locomotive coupler. A visual stain or the "bleeding out" of oil from the part would indicate a crack or other discontinuity.

The Purpose of Liquid Penetrant Testing

PT can detect discontinuities open to the surface. It is a relatively inexpensive test method considering personnel training time, materials used, and the ease with which surface discontinuities can be located. When performed properly, with proper process controls and qualified technicians, it provides reliable test results. PT is used to locate discontinuities open to the surface on the final finished surface of test objects, but is also useful for in-service checks to detect fatigue damage or production problems, for example, the root pass of some welds and fatigue cracks. Fatigue-induced cracks originate or propagate to the surface, making liquid penetrant tests useful in their detection.

Many components may fail during use. The most likely cause of material failures has been found associated with a linear-shaped discontinuity left in the test object during the manufacturing process or developed during use of the test object. This has convinced design engineers, manufacturers, users, and safety and regulatory agencies to use reliable, inexpensive surface testing to minimize the undesirable consequences and expenses of test object failure. The goal of any inspection process is to achieve as high a probability of detection as possible to mitigate risk of part failure.

Basic Principles

When a dye is added to a liquid with a certain combination of properties of cohesion, adhesion, surface tension, and viscosity, the liquid is called penetrant, penetrant dye, or in some specifications just dye. When this penetrant is placed on a clean, dry surface at the correct contact angle, it will wet the surface properly and migrate by capillary action. This capillary action causes the liquid to enter discontinuities that are open to the surface and creep up or down into the cavity.

Some specifications and company procedures require different penetrant dwell times for different types of materials or discontinuities. These times will be specifically required for the particular test object or procedure referenced. Typical penetrant dwell times are between 5 and 45 min. This penetrant dwell time allows the capillary action to cause the liquid to enter discontinuities, as shown in Figure 2. After excess surface penetrant is removed, only the penetrant in the discontinuities remains. Reverse capillary action causes the penetrant to migrate or bleed back out to form a penetrant indication.

The use of a contrasting developer aids the reverse capillary action bleedout and makes the indications easier to see. A PT indication is the penetrant bleedout spot against the contrasting developer background. After a developer dwell time, any liquid penetrant indications are evaluated to the specified acceptance criteria.

Developer dwell times vary with specification, company procedures, or techniques; type of materials; type of discontinuity to be detected; and type of developer used. These dwell times typically range between 5 min and 4 h. Figure 3 illustrates the typical sequence of a liquid penetrant test.

Introduction to Liquid Penetrant Testing

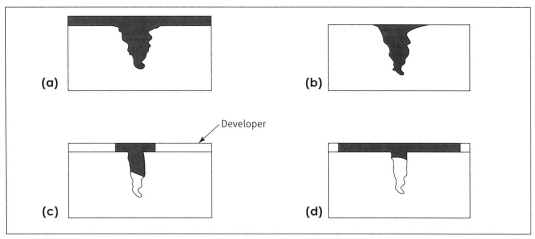

Figure 2. Penetration of liquid penetrant into a surface discontinuity: (a) applied penetrant; (b) surface penetrant removed; (c) deep, narrow crack after 30 s; and (d) deep, narrow crack after 10 min.

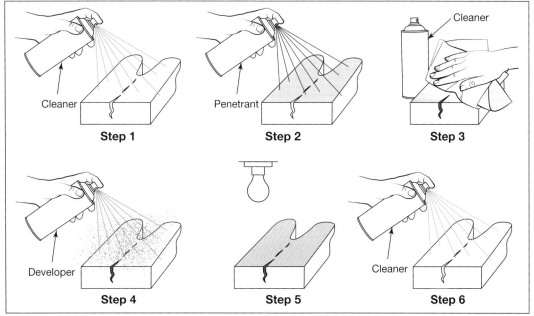

Figure 3. Sequence of liquid penetrant tests (local or spot application shown here). For process line application, immersion tanks or bulk spray applications are typically used.

- Step 1. The test object or spot is thoroughly cleaned and dried.
- Step 2. The penetrant is applied. Dwell time is allowed to penetrate discontinuities.
- Step 3. The excess surface penetrant is removed.
- Step 4. Developer is applied.
- Step 5. Indications are evaluated and accepted or rejected.
- Step 6. After the penetrant test, the test object or spot is post-cleaned.

Commercially Available Liquid Penetrant Materials

The various combinations of liquid penetrant materials that are commercially available and commonly used are changing rapidly as new and better materials are formulated. The basic classification system remains fairly uniform and classifies the penetrant dye by type: fluorescent dye, which is Type I, or visible dye, which is Type II. Type III is dual-mode penetrant that is both visible and fluorescent. The fluorescent penetrants are further classified into sensitivity levels: very low to ultra high. The different processes are then classified by the method used to remove the excess penetrant.

The developers are classified as Forms a through f. They include dry powder, water-soluble, water-suspendable, nonaqueous, and special application developers. The basic liquid penetrant classification system is shown in Table 1.

Table 1. Types and methods of liquid penetrant testing

Type	Method	Sensitivity	Developer	Solvent
Type I: fluorescent penetrant	Method A: water-washable	Level 1/2*: very low	Form a: dry powder	Class 1: halogenated
Type II: visible penetrant	Method B: lipophilic postemulsifiable	Level 1: low	Form b: water-soluble	Class 2: nonhalogenated
Type III: dual-mode penetrant	Method C: solvent removable	Level 2: medium	Form c: water-suspendable	Class 3: special application
	Method D: hydrophilic postemulsifiable	Level 3: high	Form d: nonaqueous Type I (for fluorescent)	
		Level 4: ultra high	Form e: nonaqueous Type II (for visible)	
			Form f: special application	

*sensitivity Level 1/2 applies to Type I, Method A penetrants only. There is no sensitivity level classification for Type II penetrant systems.

Personnel Qualification And Certification

The most widely used organization for NDT training and qualification guidelines is the American Society for Nondestructive Testing (ASNT). ASNT's recommended guideline is called *Recommended Practice No. SNT-TC-1A: Personnel Qualification and Certification in Nondestructive Testing*. ASNT also issues *ANSI/ASNT CP-189: Standard for Qualification and Certification of Nondestructive Testing Personnel*. ASNT certifies Level II and Level III personnel through the ASNT Central Certification Program. ASNT certifies Level III personnel who are responsible for the qualification and certification of companies' NDT Level I and Level II technicians.

When a person starts the training and qualification process, he or she is classified as a trainee. The goal is usually to become a fully certified Level II or Level III technician. Qualification is the demonstrated skill and knowledge, documented training, and experience required for personnel to properly perform the duties of a specific job. Certification is written testimony by the employer that the person is completely qualified to the company's written practice (as described in *SNT-TC-1A* or National Aerospace Standard *NAS 410, NAS Certification and Qualification of Nondestructive Test Personnel*).

To become qualified, a person must have the prescribed organized training and work experience under the supervision of a certified Level II or Level III technician. This work experience includes processing test objects as directed by the Level II technician, but never doing the evaluation for acceptance or rejection. The Level II technician supervises the processing so that if the trainee makes an error, the test can be repeated and the Level II technician can evaluate the test object and accept or reject, and the trainee will observe the evaluation.

After the company's required work experience, the trainee takes written general and specific qualification tests and a practical test on at least two test objects, or one test object for each technique to be used. If the individual passes all tests, he or she can be certified as a Level I or Level II by the company. It is possible for all the required organized classroom training and work experience to be accomplished as a trainee, and the person can be tested and certified directly to Level II.

All technicians are required to have a vision acuity test on an annual basis. This test can be taken either with or without corrective lenses. However, if corrective lenses are worn, they must be the same corrective lenses worn during PT. A color contrast differentiation examination, which can be done by utilizing the pseudo-isochromatic plates for color vision (Figure 4), must also be given to demonstrate that the individual can distinguish between the colors used for PT.

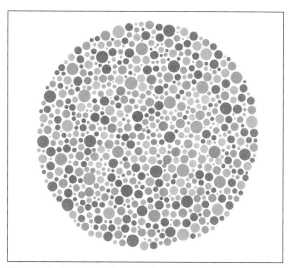

Figure 4. Sample of pseudo-isochromatic plate for color vision test.

Employee Certification

Certification comprises a combination of training and experience hours, along with various examinations to measure competency in a method.

Training
Training involves an organized program developed to provide test personnel with the knowledge and skill necessary for qualification in a specific area. This is performed in a classroom setting, where the principles and techniques of the particular test method are reviewed. The length of required training is stated in the employer's written practice.

Experience
Experience includes work activities accomplished in a particular test method under the supervision of a qualified and/or certified individual in that particular method. This includes time spent setting up tests, as well as performing calibrations, specific tests, and other related activities. Time spent in organized training programs does not count as experience. The length of experience required is stated in the employer's written practice.

Examination
Level I and Level II test candidates are given written general and specific examinations, a practical examination, and a visual examination.
The general examination covers the basic principles of the applicable method. The specific examination covers the procedures, equipment, and techniques that employees will be required to perform in their job assignment. The practical (hands-on) examination allows employees to demonstrate their ability to operate the appropriate test equipment and to perform inspections using that equipment in accordance with appropriate specifications. Level III personnel must pass written basic, method, and specific examinations as outlined in *SNT-TC-1A*. Testing requirements are stated in the employer's written practice.

Certification
Certification of NDT personnel is the responsibility of the employer. Personnel may be certified when they have completed the initial training, experience, and examination requirements described in the employer's written practice. The length of certification is stated in the employer's written practice. All applicants must have documentation that states their qualifications according to the requirements of the written practice before certifications are issued.

Safety Precautions
PT uses a variety of materials that have some hazardous characteristics. Except for water, the removers and nonaqueous carriers are usually combustible, and with unprotected contact can cause skin irritation. The developing powders are nontoxic, but can be a health hazard in confined spaces. The ultraviolet (UV) lamps used with fluorescent penetrants can cause physiological damage. All of these hazards can be avoided or minimized by observing the precautionary measures in the material safety data sheets supplied with penetrant materials.

Fire
Flashpoint is the lowest temperature at which vapors above a volatile combustible substance ignite in air when exposed to flame. The higher the flashpoint of a material, the less fire hazard it presents. Safe practice

requires that penetrant materials used in open tanks have a minimum flashpoint of 200 °F (93 °C). Smoking is forbidden in or near test areas. Penetrant materials are never stored near heat or open flame, and exhaust fans are used to disperse vapors. Aerosol cans should never be sprayed near open flames, sparks, or welding arcs.

Skin Irritation

The oil bases of liquid penetrant materials have a drying action on the skin. Because of this, the materials may cause unpleasant, if not dangerous, irritations. To prevent unnecessary contact with penetrant materials, care is taken to avoid splashing; protective hand creams are used; aprons, neoprene gloves, and face shields are worn; and soap and water are used to immediately remove any penetrant materials that have come in contact with the skin.

Air Pollution

Dust and vapors from materials used in PT are nontoxic, but inhalation of excessive amounts can be a health hazard. To avoid unhealthy concentrations of developer powder in the atmosphere, exhaust fans are installed in confined areas where dry developers are used. Fans are also used to remove the vapors from the test area. Always follow recommendations for respirator or mask use.

Ultraviolet Radiation

The UV radiation used to cause fluorescence of penetrant materials has a frequency of approximately 365 nm. This frequency is at the low end of the UV spectrum. Higher-frequency UV rays are harmful to many forms of life including humans. The UV lamp filters used with liquid penetrant tests filter out most of the harmful UV rays generated by mercury arc and light-emitting diode lamps, as well as the visible light rays.

Personnel should never look into the UV radiation. Wearing long sleeves and gloves when handling small test objects while the UV radiation source is on will help prevent any possible skin exposure. Missing, cracked, or broken filters should be replaced before use. Protective goggles or lenses should be used as required or recommended.

2

Liquid Penetrant Processing

Procedures and Techniques

All PT required by a contract or specification is performed in accordance with a referenced or supplied standard procedure or technique. Procedures can be broad and cover several specific techniques. Techniques are usually a one page data sheet that provides enough information so that any qualified Level I or Level II liquid penetrant technician can perform or repeat a liquid penetrant test on a specific test object using a standard procedure.

Each procedure and technique must be approved and signed by a certified Level III technician. In some cases, the approval certificate is submitted to the prime contractor or customer for approval, and then can be used for standard tests by a Level I or Level II certified technician.

Preparation of Test Objects

The properties of the test object determine the type of surface preparation that will be used. Nickel alloys, certain stainless steels, and titanium have an affinity for specific elements (for example, sulfur or chlorine), and if exposed to them may become structurally damaged. In all cases, the test object must be free of any foreign material that will interfere with the test and must be completely dry.

Certain cleaning and finishing operations are prohibited before liquid penetrant tests, as they may smear metal and close the surface openings of discontinuities. These operations may include power wire brushing, grit blasting, and shotpeening, but can also include other metal smearing operations.

Typical industry specifications prohibit operations that could close or mask surface openings before PT. Some specifications also require a specific technique to remove any such smeared metal, such as an approved acid etching process. A smooth surface is preferred, but PT can be satisfactorily performed on welded surfaces, relatively rough castings, and a wide variety of surface finishes using the properly designed penetrant materials and proper techniques for excess penetrant removal.

Precleaning of Test Objects

There are many satisfactory precleaning techniques depending on test material and the condition of the surface of the test object. All cleaning techniques must meet local environmental, health, and safety requirements and cause no harm to the test object. Many PT procedures have specific requirements for types of precleaning operations. As a minimum, all test objects are wiped clean with an approved solvent, and the solvent is allowed to dry completely.

Precleaning and Post-cleaning Equipment

Proper cleaning is essential to PT for two reasons:
- If the test object is not physically and chemically clean and dry, PT may be ineffective.
- If all traces of penetrant materials are not removed after the test, they may have a harmful effect after the test object is placed in service.

The cleaning processes commonly used with PT are discussed in the following sections. Unless the cleaning techniques and chemicals are known to be compatible with the penetrant, a final cleaning should be done with the solvent recommended by the manufacturer. Technicians should keep in mind that specific test objects or systems may require special cleaning materials.

Solvent Cleaning

Solvent cleaning may use tanks for immersion (Figure 1), or the solvent material may be sprayed, brushed, or wiped on and off. Solvent cleaning is the process most commonly used for spot inspections.

However, environmental, health, and safety concerns are making detergent cleaning and steam cleaning more attractive options. A solvent cleaner must evaporate readily and completely from the surface and from surface-connected discontinuities. Solvent cleaners should only be used to remove organic contaminants.

Figure 1. Small test objects immersed in acetone, a volatile cleaning agent.

Detergent Cleaning

Immersion tanks and detergent solutions are common means of performing the cleaning required by liquid penetrant tests. The detergents wet, penetrate, emulsify, and saponify (change to soap) various soils. The only special equipment requirement imposed by penetrant test cleaning is the need for suitable rinsing and drying facilities. When thoroughly rinsed and dried, detergent cleaning leaves a test surface that is physically and chemically clean. Detergent cleaners should have a combination of detergency (cleaning) dispersion, emulsifying, foaming, solubilizing, and wetting properties.

Vapor Degreasing

Vapor degreasing is effective in the removal of oil, grease, and similar organic contamination. However, there are restrictions as to its use before and after PT. Safety and environmental concerns have virtually eliminated vapor degreasing. Degreasing must be limited to those materials that have been approved for this cleaning technique. Vapor degreasing removes organic soils from both the surface and cracks, evaporates completely and, unlike water-based cleaners, does not require a rinse step or a drying (oven) step.

Steam Cleaning

Steam cleaning equipment is particularly adaptable to the cleaning of large, unwieldy test objects not easily cleaned by immersion. No special equipment is required for steam cleaning of test objects destined for PT. Steam with alkaline detergent provides ideal precleaning conditions. The alkaline detergent emulsifies, softens, or dissolves the organic contaminant, and the steam gives the mechanical action to remove the alkaline detergent/contaminant from the test object.

Ultrasonic Cleaning

Ultrasonic agitation is often combined with solvent, detergent, or alkaline cleaning to improve efficiency and reduce cleaning time. Ultrasonic cleaning equipment is useful in the cleaning of large quantities of small test objects. In many cases, special approvals must be granted to use ultrasonic agitation.

Rust and Surface Scale Removal

Any approved commercially available acid or alkaline rust remover may be used for precleaning. Required equipment and procedures are specified in the manufacturer's directions.

Paint Removal

Paint needs to be removed from a surface prior to testing (Figure 2). The dissolving type of hot tank paint strippers as well as bond release or solvent paint strippers may be used to remove paint in precleaning. The equipment and procedures required are specified in the manufacturer's directions.

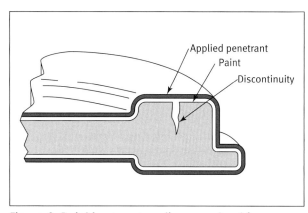

Figure 2. Paint layer preventing penetrant from entering discontinuity.

Etching

Test objects that had metal smearing operations, such as power wire brushing or sandblasting, often require etching to prepare them for PT. This process uses an acid or alkaline solution to open up grinding burrs and remove smeared metal from surface discontinuities. All acid or alkaline residues must be neutralized and removed before PT. The etching and neutralizing processes use either tanks and immersion or manual equipment and materials (Figure 3).

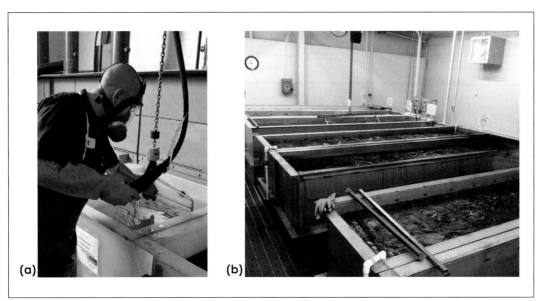

Figure 3. Prior to liquid penetrant testing: (a) etching of parts; and (b) tanks for etching aluminum.

Precleaning Processes to Avoid

Blast (shot, sand, grit, or pressure), liquid honing, emery cloth, power wire brushes, and metal scrapers should not be used on the test object before PT. These processes tend to close discontinuities by smearing metal, peening, or cold-working the surface. A hand wire brush may be helpful in removing rust, surface scale, or paint. Relatively fine bristle brushes should be used and light pressure exerted to prevent smearing of softer metals.

Drying Test Objects

A very important step in the PT process is to ensure all tested surfaces and all potential discontinuities are completely dry. If any liquid precleaner remains in discontinuities, the penetrant may be unable to enter discontinuities and an inadequate test will be performed. This is critical because the technician may not be aware that penetrant did not enter the discontinuities.

Some PT procedures require that test objects cleaned with water receive a final cleaning with a water-soluble solvent, such as isopropyl alcohol or acetone, to promote evaporation of water from discontinuities. Some PT procedures require that test objects are oven dried at temperatures up to 160 °F (71 °C). As a minimum, all surfaces should be allowed to dry by normal evaporation.

Application of Penetrant

The application of penetrant is performed after the test surface is completely clean and has been properly dried. Almost any technique is allowed for penetrant application, including spraying, brushing, pouring, or dipping. Penetrant should never be allowed to dry completely on the test surface. The test object must be turned or moved to prevent pooling of penetrant during the dwell time. It is important that all test surfaces are completely wetted with a thin coat of penetrant for the entire specified dwell time. In the fluorescent penetrant application, use of a UV radiation source will ensure that all surfaces remain wetted with penetrant and no water break areas occur where the penetrant film separates.

Dwell Time

The penetrant dwell time is the length of time that the penetrant is allowed to wet the surface and enter into discontinuities. The developer dwell time is the length of time that the penetrant is allowed to absorb into the developer before the indications are evaluated.

Removal of Excess Surface Penetrant

The removal of excess surface penetrant is an integral step in the PT process and requires proper lighting and rinse conditions.

Adequate Illumination

Adequate illumination at the rinse station for fluorescent penetrant is no more than 10 fc (100 lx) of white light and greater than 100 $\mu W/cm^2$ of near UV radiation (UV-A) unless stated otherwise in an industry specification. This lighting allows better monitoring of removal of excess fluorescent penetrant and limits possible over rinsing or under rinsing.

The UV radiation background lighting at the rinse station also allows the technician to ensure complete coverage of the test object when using an emulsifier, since the emulsifier is a different color than the penetrant under the UV radiation. The rinse station may be illuminated to the same level as the inspection booth (1000 μ/cm^2 and less than 2 fc [20 lx]), if desired.

The evaluation of liquid penetrant indications requires proper illumination of the complete testing area. This is usually specified in the PT procedure. For visible dye penetrant, 100 fc (1000 lx) at the test surface is adequate. For fluorescent penetrant tests, the surface should be illuminated with 1000 $\mu W/cm^2$ of UV-A as measured at the test surface, and the white light should be less than 2 fc (20 lx). These lighting levels will meet the requirements of most industrial specifications. However, verification of the proper illumination requirement should be given in the standard procedure or technique being used.

Personnel in the testing area should not wear white clothing, which reflects large amounts of visible light when exposed to UV radiation.

Water Rinse (Methods A, B, and D)

After the required penetrant dwell time and proper emulsification (as applicable), the water rinse should be coarse droplets normally applied at an oblique angle from a distance of approximately 12 in. (30 cm), as shown in Figure 4. The standard nozzles available from approved NDT suppliers have been qualified for standard industry techniques. The preset pressure should not exceed 40 psi (275 kPa), and the temperature should be between 50 and 100 °F (10 and 38 °C). When hydro-air nozzles are used, the air pressure should not exceed 25 psi (172 kPa).

Figure 4. Fluorescent penetrant about to be rinsed from test parts following dwell time.

The rinse is typically accomplished at a rinse station with adequate illumination (UV radiation for Type I penetrant and white light for Type II penetrant). The light should illuminate the test object so that the technician can evaluate when the excess penetrant is removed. The technician should also ensure there is no overwashing. When the excess surface penetrant is removed, the surface water is drained or removed from cavities, holes, or pockets. If allowed by the procedure, blotting or even a filtered air spray at less than 25 psi (172 kPa) may be used, but care should be used not to smear any indications that may begin to bleed out.

The drying technique will depend on the type of developer used and the test object. The drying time should be the minimum required so that development and evaluation can be performed as quickly as possible.

Manual Wipe (Method C)

After the required penetrant dwell time, the test object is first wiped with a clean, dry, lint-free cloth, or absorbent towel (Figure 5). The towel should be white, or a color that contrasts with the penetrant. The removal of the bulk excess penetrant works best if clean sections of cloth are used for each wipe.

Figure 5. Removal of excess penetrant by wiping with a soft, dry cloth.

The removal process should start from the most difficult areas to clean, such as areas of acceptable weld undercut or ripples. After the bulk of the excess penetrant is removed with dry cloths, the remaining penetrant is removed with dampened cloths. In no case should the cloths be saturated with cleaner. If any cleaner drips or can be squeezed from the cloth, it is too wet and very shallow indications may be removed.

When the surface of the test object is visually free from penetrant and the wipe cloth is coming out relatively clean, the wiping process is complete. The technician then watches the solvent or water evaporate until the surface is dry, and then the required type of developer may be applied. Some procedures allow a dryer oven with recirculating air, typically not warmer than 160 °F (71 °C).

Developer Application and Drying

There are four basic types of developers:
- dry powder (fine powder form)
- water-soluble (liquid dip tanks)
- water-suspendable (liquid dip tanks)
- nonaqueous (aerosol spray can)

All developers must be applied so that a thin uniform coating covers the entire test surface. Application of an excessive thickness can obscure, cover, or extinguish both visible and fluorescent indications. The classifications and proper application are as follows.

Dry Powder

Dry developer is loose, fluffy powder used with fluorescent penetrants. After removal of excess penetrant and/or emulsification and drying, dry powder developer is applied to the test surface for the purpose of absorbing penetrant from discontinuities and enhancing the resultant penetrant indications.

Dry developer may be applied in a dust cloud chamber activated by an air blast. The developer can be applied with an electrostatic sprayer (Figure 6), or the technician can dip the test object into the dry powder. Excess powder is shaken or tapped off of the test object. In some cases light air blow-off not exceeding 5 psi (34 kPa) is an acceptable alternative.

Figure 6. Electrostatic application of dry developer to a titanium test object.

Of the different developers available, dry developer is the most adaptable to rough surfaces and automatic processing. It is also the easiest to handle, apply, and remove. Dry powder is not corrosive, gives off no vapors, and leaves no residue or film to affect the next processing step. Dry powder sensitivity is slightly lower than water-soluble for smooth test objects, but it is more sensitive for rough test objects or threads. Dry developer should not be used with visible dye penetrant because of the poor contrast provided by the thin coating of very fine powder.

Water-soluble

Water-soluble developers are made from powder crystals that dissolve in a solution mixed with water, following the manufacturer's recommendations. The developer solution concentration is verified by checking the specific gravity of the liquid with a hydrometer after mixing. Water-soluble developer concentration will change over time due to evaporation losses. Therefore, the developer concentration should be checked on a weekly basis. The application occurs immediately following excess penetrant removal and/or emulsification from the test object and after draining or shaking off excess water, but before drying time.

Wet developer is applied by dipping (immersion), flow, or spray techniques. Immersion in a prepared tank of developer is the most common application technique. With immersion testing, the test object is immersed only long enough to coat all surfaces. It should then be removed immediately, and all excess from recesses or trapped areas should be drained to prevent pooling of developer, which can obscure indications. Wet developer is applied to form a smooth, even coating.

Particular care should be taken to avoid concentrations of developer in dished or hollowed areas of the test object. Such concentrations of developer may mask penetrant indications and are to be avoided. This developer is not normally used for critical applications on complex geometry test objects.

Water-suspendable
The water-suspendable application is a suspension of white powder mixed with water, following the manufacturer's recommendations. Unlike water-soluble developers (which do not require agitation), water-suspendable developers require constant mild agitation (or thorough agitation before and during use) to keep the powder particles in suspension.

The water-suspendable mixture concentration is verified by checking the specific gravity with a hydrometer after mixing. Water-suspendable developer concentration will change over time due to evaporation losses. Therefore, the developer concentration should be checked on a weekly basis. The application immediately follows excess penetrant removal from the test object and after draining or shaking off excess water, but before drying time.

Water-suspendable developer is applied the same as water-soluble developer. In addition, the same care must be taken in the application of the developer.

The fact that constant agitation is required to maintain the proper mixture/concentration is a disadvantage of water-suspendable developers. Care should be taken not to generate foam during agitation because foam will cause uneven surface coating. As with water-soluble, this developer is not normally used for critical applications on complex geometry test objects.

Nonaqueous
Nonaqueous wet developer is a powder suspended in a volatile liquid solvent. The volatile liquid gives the advantage of decreasing the viscosity and increasing the liquid bulk in the cavity. This action forces the penetrant to the surface where it assists the dispersion of the penetrant away from the discontinuity. The evaporation of the solvent tends to pull penetrant into the developer.

Nonaqueous wet developer is the most sensitive developer. The application is by spraying, and the volatile liquid evaporates rapidly so that no delayed drying operation is required. The spray or aerosol container must be thoroughly agitated just before spraying, and spraying should be done sparingly so that a thin coating covers the entire test area. The sheen of the metal should barely be covered.

Several very light coats are usually preferable, rather than attempting to cover the test area with one spray. A light check spray should be performed away from the test surface to ensure the spray nozzle is clean and free of obstructions.

Development Time

Proper development time is very important in the interpretation and evaluation of indications. Development time is the time allowed between the application of developer and the viewing of indications. If insufficient time is allowed, indications may not have time to develop. Too much time will cause indications to become blurred or distorted.

The required interpretation time depends on the developer, and is stated in the procedure or specification. With a nonaqueous wet developer, the normal time until evaluation is from 10 min to 1 h. With an aqueous wet developer, the normal time until evaluation is 10 min to 2 h. With a dry developer, the normal time until evaluation is from 10 min to 4 h.

Interpretation and Evaluation

The terms interpretation and evaluation are often confused. Actually, the terms refer to two entirely different steps in the testing process. The success and reliability of the interpretation and evaluation of PT indications is dependent on the thoroughness, training, and dedication of the technician. PT is not a method by which a test object is processed through a machine that separates the good test object from the bad. Testing personnel are required to carefully process each test object, interpret indications, and then evaluate the indication to the acceptance standard referenced or given in the procedure they are performing.

Interpretation

To interpret an indication is to first decide if it is false, relevant, or nonrelevant, and determine what condition caused it. One technique of determining whether an indication is relevant is to dampen a cotton swab with solvent and gently wipe the indication off. If the indication reappears, it is a relevant indication. This technique is known as the bleed-back evaluation technique, and should not be repeated more than twice. If the indication is a false indication, the technician must determine the source of the false indication and correct the problem. Test objects with false indications must be recleaned and reprocessed if the false indications interfere with the detection of relevant indications.

Valid indications may be relevant or nonrelevant. Relevant indications are caused by a surface discontinuity in the test object. Nonrelevant indications are caused by a design feature, such as a press fit or a rivet. When the technician determines that the indication is relevant, it must then be evaluated to the acceptance criteria.

Evaluation

To evaluate an indication means to decide if it is acceptable, requires rework, or causes the test object to be rejected. Most industry acceptance standards are written with clear specifications. These acceptance standards include acceptance limits, which are stated in terms of size, location, and proximity to one another and specify types of discontinuities. For some applications of PT, a technician may only be told that no linear cracks or no linear indications are allowed.

When inspecting for material discontinuities, the technician determines if the indication is a crack, seam, lap, porosity, lack of fusion, or hot tear. This determination depends on the knowledge of the test object's history, and the company policy for evaluation. In some cases, a metallurgist may be asked to determine the cause of the discontinuity.

The next step may be to measure each discontinuity, depending on the acceptance criteria, and fill out a detailed report, including a sketch or photograph to submit for further engineering evaluation. In many cases, a Level II technician evaluates and either accepts or rejects the test object. It is very important to mark and separate rejected test objects.

Indications

A liquid penetrant indication is caused by a discontinuity on the surface or by penetrant remaining on the surface from a false or nonrelevant cause. However, proper interpretation requires familiarity with the manufacturing processes, the types of discontinuities likely to occur, and their appearance. It requires knowledge of how different materials fail and where failure occurs. It also requires the careful control of the penetrant process. The effects of these variables are discussed in the following sections. Please refer to Chapter 6 for detailed photographs that illustrate typical indications found in PT.

False Indications

The most common source of false indications is poor cleaning of test objects and inadequate removal of excess penetrant. UV radiation is very important during the washing process when using fluorescent penetrant. The technician determines whether a good rinse is obtained or whether patches of fluorescence remain on the test object. Adequate lighting at a fluorescent rinse or cleaning station is usually considered less than 10 fc (100 lx) of white light and 100 µW/cm^2 of UV radiation at the test object surface.

For solvent-removable penetrants, the technician controls the removal process. This allows for a slower and more thorough test. To guard against confusion resulting from fluorescent or color spots other than relevant indications, care is taken so that no outside contamination occurs. Typical sources of contamination include the following:
- penetrant on the hands of the technician
- contamination of wet or dry developer
- penetrant rubbing off of an indication on one test object to a clean portion of the surface of another test object
- penetrant spots on the testing table

To avoid contamination, process tanks and testing areas are kept clean. Only lint-free wiping cloths or rags are used, and test objects are kept free of fingerprints and tool marks. Figure 7 illustrates common types of false indications, caused by handling or cleaning processes.

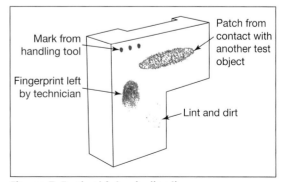

Figure 7. Typical false indications.

Nonrelevant Indications

Nonrelevant indications are caused by features in the test object that are there by design, but are in no way a relevant discontinuity. Most nonrelevant indications are easy to recognize because they are related directly to some feature of the assembly that accounts for their presence.

Nonrelevant indications include those that appear on test objects that are press fitted, riveted, or spot welded together, and those caused by surface roughness.

Any nonrelevant indication that interferes with or could mask a relevant indication must be further evaluated and sometimes retested. If a test object is too rough to perform a proper liquid penetrant test, or has a condition like weld undercut that has been accepted visually, the test object or weld may have to be returned for further preparation for PT.

Relevant Indications

Relevant indications are those caused by a material discontinuity. The interpretation of an indication as relevant is a matter of observing the indication, eliminating the possibility of it being a false indication and then further determining that it is not a nonrelevant indication. Any relevant indication immediately becomes subject to evaluation of its cause (type of discontinuity).

Some procedures allow measuring the visible discontinuity size after wiping off the developer. This should be done very carefully, with magnification and reapplication of developer to compare the indication size measured visually to the penetrant indication. For example, a small diameter, deep, visible porosity may bleed out as a very large indication after the minimum dwell time. This fact should be reported.

Categories of Relevant Indications

Discontinuity indications vary widely in appearance, but for each indication two interpretive questions must be answered.
- What type of discontinuity caused the indication?
- What is the severity of the discontinuity as evidenced by the extent of the indication?

The answers to these interpretive questions are obtained by observing the indication and identifying the discontinuity by evaluating the characteristic appearance of the indication.

Each indication may also require an answer to the evaluation question, what effect will the indicated discontinuity have on the service life of the test object? The answer to the evaluation question is based on certain knowledge of the seriousness of the discontinuity and complete understanding of the test object.

Figure 8 illustrates typical relevant indications (more examples of relevant indications can be found in Chapter 6). Relevant indications fall into five categories:
- continuous linear indication
- intermittent or broken line indication
- rounded indication
- faint indication
- gross indication

Liquid Penetrant Processing

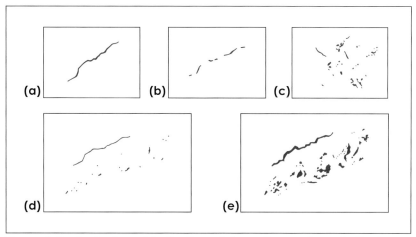

Figure 8. Typical relevant indications: (a) continuous linear indication; (b) intermittent or broken line indication; (c) small dots and rounded indication; (d) faint indication; and (e) gross indication.

Continuous Linear
Continuous linear indications are caused by cracks, seams, cold shuts, forging laps, scratches, or die marks. Linear indications are those that are three times greater in length than width. Cracks usually are linear and irregular, rather than straight. Seams appear as straight or spiral lines. Cold shuts appear as smooth, narrow, regular solidification lines. Forging laps appear as smooth lines. Scratches and die marks appear in a variety of linear patterns, but are recognizable when all penetrant traces are removed since the bottom of the discontinuity is usually visible. Scratches, die marks, and other surface discontinuities are stress risers, and some test objects require evaluation of these conditions.

Intermittent Linear
The same discontinuities that cause continuous linear indications may, under different circumstances, cause intermittent linear indications. When a test object is worked by grinding, peening, forging, or machining, portions of the discontinuities on the surface of the test object may be closed or removed by the metal working process. When this occurs, the discontinuities will appear as intermittent lines. Machining operations may also open linear indications that lie below the surface, exposing only the tops of continuous indications and thus making them appear as intermittent indications. Some acceptance standards consider separation of indications and require evaluation and reporting of the total length to the extremities of the indications. Most specifications require some type of etch to be used if metal smearing has occurred.

Rounded
Rounded indications are usually caused by porosity. The porosity may be the result of gas holes, weld porosity, or the generally porous makeup of the test object. Rounded indications are those that are less than three times greater in length than width. Deep cracks (especially weld crater cracks) may also appear as larger rounded indications after they bleed

out because they trap a large amount of penetrant that spreads when the developer is applied.

Any large, rounded liquid penetrant indication may appear as an elongated porosity on a radiograph, and may have some length if tested by ultrasonic testing. After evaluation, a technician may wipe away a large, rounded indication and reapply developer, then watch the formation during the bleedout to obtain more information about the nature of the discontinuity.

Faint
Faint indications are particularly difficult to interpret. Faint indications appearing over a large area are always suspect. When they appear, the test object should be thoroughly cleaned and retested. If the indications were not caused by inadequate cleaning, the test object may require more preparation.

Gross
A gross indication is one that stands out without the need for penetrant. In other words, it is an indication that is so severe that it can be seen with the naked eye.

Discontinuity Depth Determination
When evaluation requires more accurate knowledge of the relative depth of a discontinuity, it may be obtained by removing the surface indication and reapplying developer. If the subsequent amount and rate of penetrant bleedout is fast, large, and bright, then it is a deep discontinuity. This is why many acceptance standards reject large, rounded indications. There is no way to estimate a dimension for the depth of a discontinuity with PT.

Post-cleaning
If all traces of penetrant testing materials are not removed after test, they may have a harmful effect when the test object is placed in service. The cleaning processes typically used with PT are discussed in the following chapters. In some cases, the preparation for the next production process, such as plating or painting, may be sufficient. Components should be cleaned after examination to remove developers and other examination material residues if these are detrimental to subsequent operations or to the test object's intended function.

Post-cleaning Materials
Following the test, the residue of penetrant materials may need to be removed. Post-cleaning is particularly important when test objects are destined for use in a liquid oxygen environment. In fact, liquid oxygen-compatible penetrant materials must be used to test these objects. Though many test objects will receive further processing, the cleanliness of any test object after completion of a penetrant test is typically the responsibility of test personnel.

Except for liquid oxygen compatibility and the chlorine-free requirement in the precleaning and post-cleaning of nickel alloys, certain stainless steels, and titanium, no special materials are required for post-cleaning unless required by specification or company procedures.

Liquid Penetrant Testing Methods

Standard Methods

There are several organizations that issue standards for PT. There are minor differences in these standards, which are easily incorporated into a single liquid penetrant procedure that covers the Type I (fluorescent), Type II (visible), and Type III (dual-mode) processes and four methods of PT (see Table 1 in Chapter 1). Each company performing PT is required to have its procedures approved for the basic standard requirements.

Three of the more common standards used throughout the PT industry are:

- *SAE AMS 2644, Inspection Material, Penetrant: an Aerospace Material Specification*: published by SAE International, which establishes the classification, technical requirements, tests, and test procedures for the qualification, approval, and quality verification of all materials used in PT.
- *ASTM E1417, Standard Practice for Liquid Penetrant Testing*: an ASTM International practice that establishes the minimum requirements for conducting PT of nonporous metal and nonmetal components. This standard covers general and specific practices along with process control requirements.
- *ASTM E165, Standard Practice for Liquid Penetrant Examination for General Industry*: a practice that covers procedures for penetrant examination of materials including a basic summary of the inspection process, the significance and use of liquid penetrant, how penetrant is classified, the materials that are used during PT, specific procedures, special requirements when performing liquid penetrant inspections on certain materials, qualification and requalification of inspection personnel, and cleaning of parts and materials.

Method Characteristics

Making decisions about which liquid penetrant to use requires knowledge of the characteristics of each type. Knowing these characteristics will ensure that the correct material selection is made.

Fluorescent or Visible

The choice to use Type I (fluorescent), Type II (visible), or Type III (dual-mode) penetrant is sometimes based solely on the availability of a dark area and power for the UV radiation, but is more often based on the necessity of the specified sensitivity levels available only with

Type I (fluorescent) penetrant processing. All the methods require the same thorough precleaning and drying before penetrant application.

The greatest difference between Type I and Type II systems is the sensitivity rating system for Type I. Type I systems are generally more sensitive than Type II systems. Because of this greater sensitivity, the use of Type I systems is required on most critical aerospace components. Type II penetrant should never be used before Type I penetrant, because any red penetrant remaining in discontinuities will dilute the fluorescent penetrant and the fluorescent indication may not appear under UV radiation.

Solvent- or Water-removable

The portable solvent- or manual water-removable penetrant kits allow field tests at construction sites, aboard ships, or on large assemblies in fabrication shops. The materials are in aerosol spray cans. This allows one technician to carry all the equipment and perform the test. The Type II (visible), Method C (solvent-removable) test is the most common portable method, but in some cases water-removable penetrant techniques can be used. Portable fluorescent kits are also available. Aerosol cans do not require daily checks or contamination checks.

Inline Penetrant Systems

Inline penetrant systems consist of bulk penetrant materials in tanks (Figure 1). Type I (fluorescent) is preferred because the indications are easier to see. The size and design of the penetrant line depends on the type and size of the test objects. The tanks can be very small in a separate room, or built so the entire assembly line passes through the penetrant process. The manual water rinse is the most common penetrant removal method. Emulsification dip tanks or sprayers may be used if required for certain test objects. Inline penetrant systems are used to test a large volume of manufactured or cast test objects, but are also commonly used for overhaul and maintenance shops.

Figure 1. Bearings after penetrant application using an inline Type I (fluorescent) liquid penetrant testing system.

Method A (water-washable) is the most common penetrant method. Method A penetrant is normally sensitivity Level 2 or 3, but sensitivity Levels 1/2, 1, 2, 3, and 4 are available. The test object and specification requirement will determine which sensitivity level should be used. The clean, dry test objects are typically dipped briefly in the tank of penetrant, then moved to a drain station that allows the excess penetrant to drain back into the penetrant tank during the dwell time.

After the required dwell time, the test objects are taken to the rinse station, which has appropriate lighting. The background fluorescent lighting makes it easy to determine when the excess penetrant is removed. If nonaqueous or dry developer is used, the test object is then ready for the drying oven. If aqueous developer is used, the test object is then dipped in wet developer.

The postemulsified Methods B or D can also be added to a liquid penetrant line by adding the emulsifier dip tank(s) or spray equipment. The postemulsified methods are normally used when very shallow discontinuities are critical, or when every possibility of eliminating over rinsing of test objects is a requirement. Critical objects, such as turbines or aircraft engine parts, often require high sensitivity postemulsification methods. Special restrictions and requirements may be specified.

Method A, Water-washable

Method A, water-washable (Figure 2) is the simplest and most widely used penetrant testing method and is normally medium or high sensitivity. Typically, the method of application is to immerse the clean/dry part into a tank full of penetrant material, then move the test part to a drain station. A dwell is implemented to allow the penetrant material to propagate into existing cracks and other fissures. After a required dwell time, penetrant on the surface is removed with a coarse water spray. The penetrant itself contains an emulsifying agent (or detergent). This self-emulsifying penetrant allows for washing the penetrant from the surface of test with just the use of water. The mechanical force of the water spray converts the penetrant into small, suspended oil droplets. After the water spray, a developer is applied to the test surface. The developer helps to draw out remaining penetrant material that

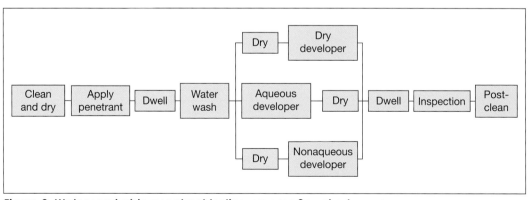

Figure 2. Water-washable penetrant testing process flowchart.

has propagated into cracks and other fissures in the test part. In some cases a portable water-removable penetrant technique can be used to inspect parts that are not appropriate to send through a penetrant line. More information on the water-washable penetrant testing method can be found in *ASTM E1209: Standard Practice for Fluorescent Liquid Penetrant Testing using the Water-Washable Process.*

Method B, Lipophilic Emulsification

Method B is the lipophilic (oil-based) emulsification method (Figures 3 and 4). After the proper oil-based penetrant is applied and the dwell time has elapsed, the test object is ready for emulsification. The lipophilic emulsifier is typically located in a dip tank, and is a contrasting color from the penetrant so that it forms a visible coat over the penetrant to ensure complete coverage.

Application is typically done by immersion, but flowing may be used. No agitation of the test object is allowed. The lipophilic emulsifier acts by diffusing into the excess oil base surface penetrant and making it water-washable.

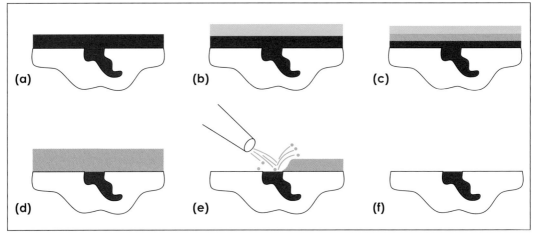

Figure 3. Acting mechanism of lipophilic emulsifiers: (a) apply penetrant; (b) apply emulsifier; (c) diffusion begins; (d) diffusion complete; (e) rinse; and (f) clean surface.

The emulsification time is determined by experimentation and depends on the features of the test object, the uniform dipping and draining of the test object, and the viscosity of the emulsifier. This makes uniform dipping, draining, and emulsification time very important, so this method is typically monitored in seconds. The maximum time allowed is 180 s for Type I systems and 30 s for Type II systems, and will be shorter for smoother test objects. After the proper emulsification, the water rinse is identical to the Method A rinse and developing will be the same as other methods.

Liquid Penetrant Testing Methods

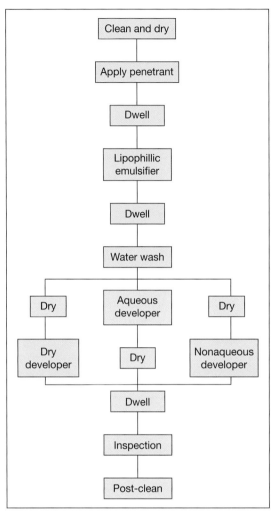

Figure 4. Postemulsifiable lipophilic penetrant testing process flowchart.

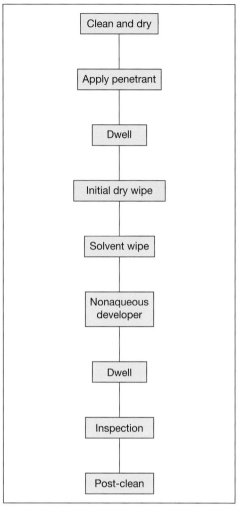

Figure 5. Solvent-removable penetrant testing process flowchart.

Method C, Solvent-removable

These penetrants do not use an emulsifier to remove penetrant from the surface of a test object. Instead, a solvent wipe is incorporated to remove excessive penetrant from the test part (Figure 5). Typically, the removal of penetrant is accomplished by first wiping the surface of the test piece with a clean rag, and then performing a second removal with a solvent-dampened rag until background fluoresence is reduced to a usable level. Solvent-removable penetrant is most commonly incorporated through a portable process and not in a penetrant line. The penetrant material for this method is typically stored in an aerosol spray can, making this a common portable testing method. An advantage of this method is the lack of frequent process controls incorporated, which are necessary for other methods to ensure the process performs adequately.

29

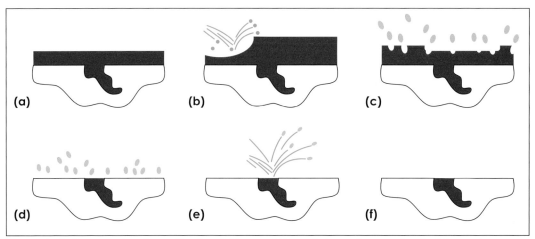

Figure 6. Acting mechanism of hydrophilic emulsifiers: (a) apply penetrant; (b) pre-rinse; (c) detergent action begins; (d) detergent action ends; (e) rinse; and (f) clean surface.

Method D, Hydrophilic Emulsification

Method D is the hydrophilic (water-based) emulsification method (Figures 6 and 7). The test object is ready for emulsification after the proper oil-based penetrant is applied and the dwell time has elapsed. The hydrophilic emulsifier can also be used in tanks, or spray equipment can be modified so a mix of hydrophilic emulsifier and water can be used to remove the oil-based penetrant.

Hydrophilic emulsifier is water-based and is supplied in a concentrated form that is diluted in water to concentrations of 10 to 30% for dip applications, and 0.05 to 5% for spray applications. A water pre-rinse is required to help remove some of the bulk penetrant before immersing in the emulsifier. This helps reduce some of the penetrant contamination in the emulsifier tank.

Hydrophilic emulsifier acts on the penetrant from the surface by detergent action. The spray or agitation in the tank provides a scrubbing action.

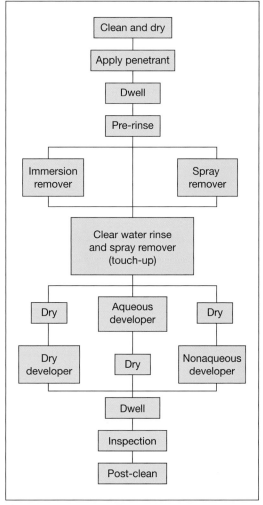

Figure 7. Postemulsifiable hydrophilic penetrant testing process flowchart.

The emulsification time for dip tank applications is determined by experimentation and is normally 120 s maximum.

The manufacturer will specify the proper concentration of emulsifier in water, which should be checked periodically with a refractometer.

Tanks of emulsifier will mix with a small amount of penetrant during the dipping process over a period of time. This is one reason for periodic checks to monitor the system materials and performance.

Liquid Penetrant Testing Equipment

Liquid Penetrant Testing Units

PT units can be arranged so test objects are moved manually, mechanically assisted, semiautomatically, or fully automatically. This choice depends on budget, timing, and long-term system performance. The size of the testing unit is largely dependent on the size and types of the test objects. The layout of the equipment may be U-shaped, L-shaped, or a straight line, and is determined by the facilities available, production rate, and required ease of handling.

Stations

Depending on the type of penetrant and processing used, PT units require stations, as shown in Figures 1 through 5. In a typical testing facility for a postemulsification process, the following stations are required:
1. Precleaning station (usually remote from penetrant test station)
2. Penetrant station (tank, tunnel, or application booth)
3. Drain station (used with penetrant tank)
4. Pre-rinse station
5. Emulsifier station (tank, tunnel, or application booth)
6. Rinse station (sink with lighting)
7. Developer station (tank, tunnel, or application booth oven)
8. Dryer station (usually an oven)
9. Inspection station (enclosed booth or table with proper lighting)
10. Post-cleaning station (usually remote from the penetrant test station)

Auxiliary Equipment

Auxiliary equipment is defined as the equipment located at penetrant test stations (other than cleaning stations) required to perform PT. In some instances, the auxiliary equipment may be built in at one or more of the test stations. Auxiliary equipment includes light sources, refractometers, and timers.

Pumps

Various pumps installed at the penetrant, emulsifier, rinse, and developer stations are used to agitate the solutions, pump drain off material into the proper tank for reuse or filtration for disposal, and power handheld sprayers and applicators.

Liquid Penetrant Testing Classroom Training Book | CHAPTER 4

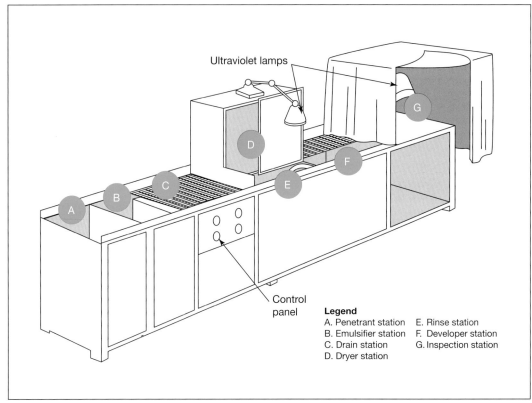

Figure 1. Typical small test equipment using fluorescent postemulsifiable penetrant and dry developer.

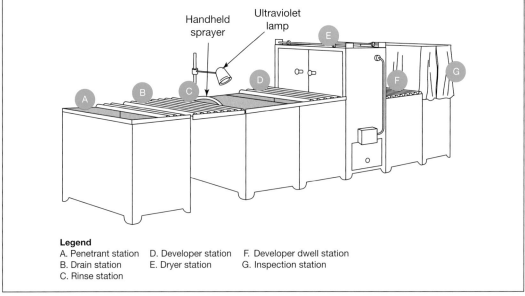

Figure 2. Typical medium test equipment using fluorescent water-washable penetrant and wet developer.

Liquid Penetrant Testing Equipment

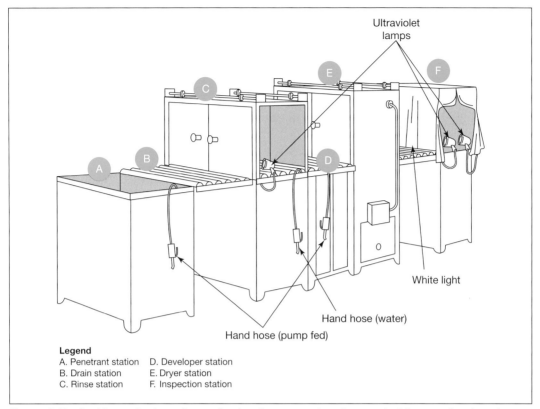

Figure 3. Typical large test equipment using fluorescent water-washable penetrant and wet developer.

Figure 4. Typical fluorescent penetrant system with test pieces processed counterclockwise: water-washable penetrant dip or dwell tanks; wash station; test piece drying ovens; developer dust chamber; darkened inspection room (not shown) at end of line.

Figure 5. Automated liquid penetrant testing equipment.

Sprayers and Applicators

Sprayers and applicators are frequently used at the penetrant, emulsifier, rinse, and developer stations. They decrease test time by permitting rapid and even application of penetrant materials. Both conventional and electrostatic sprayers are used for penetrants, dry developers, and suspendable developers. Examples of electrostatic spray systems are shown in Figure 6.

Lamps

White lamps, as well as UV lamps, are installed as required to ensure adequate and correct lighting at all stations. When fluorescent materials are used, UV lamps are installed at both the rinse and test stations.

Timers

One or more timers with alarms are used to control penetrant, emulsifier, developing, and drying cycles.

Thermostats and Thermometers

Thermostats and thermometers are used to control the temperature of the drying oven and penetrant materials.

Exhaust Fans

Exhaust fans are used when testing is performed in closed areas. The fans facilitate removal of fumes and dust.

Refractometers

A refractometer (Figure 7) is used to measure the concentration of hydrophilic emulsifier in water.

Hydrometers

The hydrometers used in PT are floating instruments, as shown in Figure 8. They are used to measure the specific gravity of water-based wet developers.

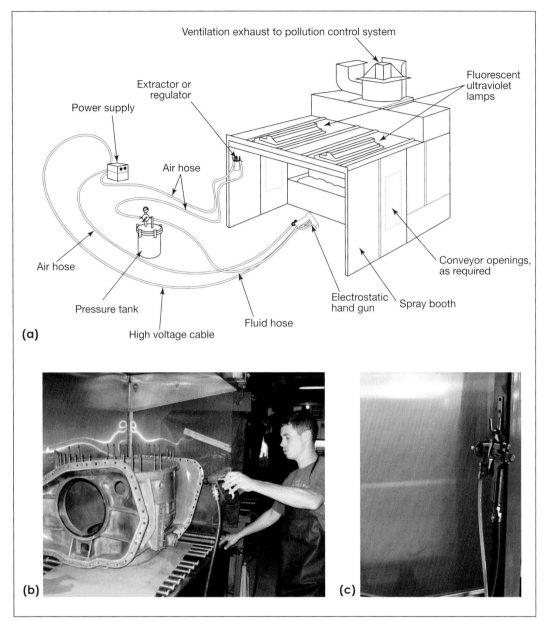

Figure 6. Application with an electrostatic spray system minimizes consumption of liquid penetrant material and helps ensure coverage of complete surface: (a) diagram of electrostatic spray system components; (b) a technician applying fluorescent liquid penetrant with a handheld electrostatic spray gun that makes atomized particles attract to the test object; and (c) automatic electrostatic spray.

Figure 7. Typical refractometer.

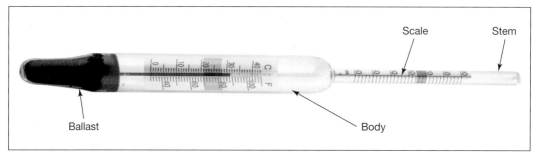

Figure 8. Typical hydrometer (shown horizontally, but used vertically).

Portable Equipment

It is possible to perform PT without stationary equipment. When testing is required at a location remote from stationary equipment, or when only a small portion of a large test object requires testing, portable liquid penetrant kits may be used. Both fluorescent and visible dye penetrants are available in kits. The penetrant materials are usually dispensed from pressurized spray cans or the penetrant may be applied with a brush.

Visible Dye Penetrant Kit

The visible dye penetrant test kit is lightweight and contains the materials necessary for a test, as shown in Figure 9. It consists of a container with at least the following items:
- solvent cleaner
- penetrant remover
- visible penetrant
- application brushes or pads
- nonaqueous wet developer
- wiping cloths

Liquid Penetrant Testing Equipment

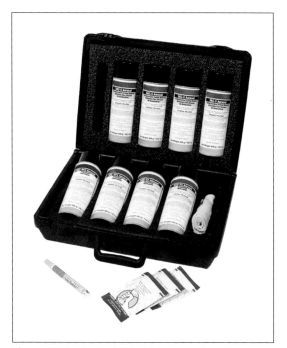

Figure 9. Typical visible dye penetrant portable kit.

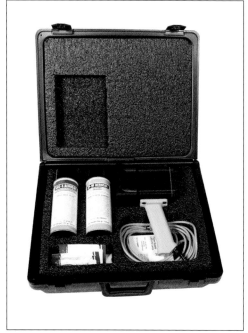

Figure 10. Typical fluorescent penetrant portable kit.

Fluorescent Penetrant Kit

The fluorescent penetrant kit combines portability with the high visibility associated with fluorescent materials. The kit holds all the essential materials required for the test, including a UV lamp, as shown in Figure 10. The fluorescent kit consists of a container with the following items:
- portable UV lamp
- solvent cleaner
- penetrant remover
- fluorescent penetrant
- application brushes or pads
- developer, either nonaqueous wet or dry
- wiping cloths
- hood to provide a darkened area for viewing indications

Ultraviolet Radiation

The electromagnetic spectrum including the UV radiation used for evaluation of fluorescent liquid penetrant indications is shown in Figure 11. Handheld UV mercury vapor arc lamps (Figure 12) produce filtered light in the wavelength range of 320 to 440 nm. UV mercury vapor arc lamps cause the fluorescent penetrant to fluoresce visible light in the green-yellow or red-orange wavelengths (between 520 and 580 nm), depending on the material being viewed. The equipment usually consists of a mercury arc bulb and a filter. The bulb and filter are contained in a reflector lamp unit.

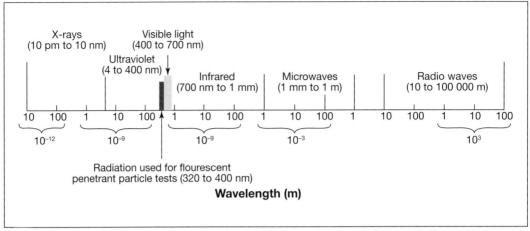

Figure 11. Electromagnetic spectrum showing the narrow range of UV radiation used for fluorescent PT.

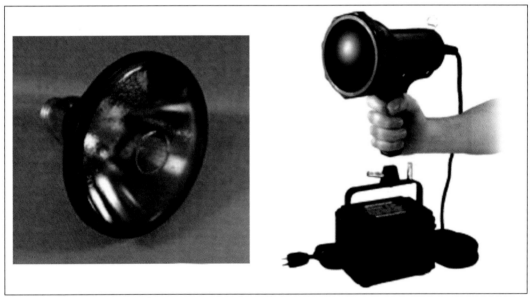

Figure 12. Mercury vapor lamp.

    For correct test results, the lamp should produce an intensity of at least 800 µW/cm² at the test surface, and most specifications require the output to be 1000 µW/cm² measured at 15 in. (38 cm), which should be checked on a daily or weekly basis. This equipment check is needed because the lamp loses its intensity over a period of usage and will have to be replaced before it burns out.

    The lamp's deep red-purple filter is designed to pass only those wavelengths of light that will activate the fluorescent penetrant material, which is approximately 365 nm. It also helps filter out harmful UV radiation. Because dust, dirt, and oil greatly reduce the intensity of the emitted light, the filter should be frequently cleaned.

When using mercury vapor lamps, the full intensity of the lamp is not attained until the mercury arc is sufficiently heated. At least a 5 min warm-up is required to reach the required arc temperature.

Because switching the lamp on and off shortens bulb life, the lamp is usually left on during the entire test period. If the UV radiation is switched off, it will not respond immediately when turned on. It may take approximately 10 min to reach full intensity again.

Fluorescent UV radiation tubes are also available that emit 320 to 400 nm wavelengths, but focus is difficult and these bulbs are usually not adequate for evaluation stations. They are most commonly used for rinse stations or auxiliary lighting.

Light-emitting diode (LED) lamps (Figures 13 through 15) are becoming more popular, with many different models currently available.

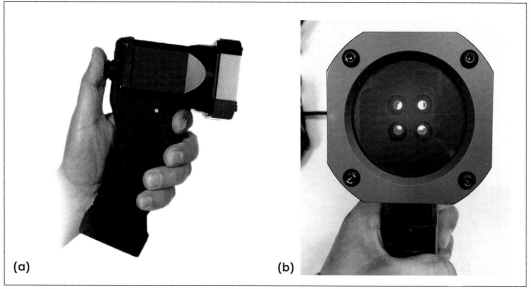

Figure 13. Battery-powered UV-A LED lamps: (a) side view; and (b) front view.

Figure 14. UV-A LED lamp on a fluorescent liquid penetrant.

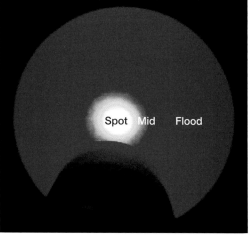

Figure 15. Typical beam spread coverage of a UV-A LED flood model lamp.

LED lamps are available as portable, handheld devices or stationary models for designated inspection areas. LED UV-A lamps are generally more durable (in comparison to 100 W bulbs) and capable of providing high intensity output levels. Some industry specifications prohibit intensities greater than 5000 µW/cm². Governing specifications may call for additional requirements during the selection of LED UV-A equipment.

Tubular Fluorescent Cold Discharge Sources

Electrically and mechanically, these standard fluorescent bulbs come in sizes from 2 W to more than 60 W input. These cold discharge tubular lamps contain low-pressure mercury vapor glow discharges. Their primary radiation is hard UV radiation of 253.7 nm wavelength. This is used to excite a special cerium-activated calcium phosphate phosphor coated on the inside of the tube. This phosphor, when activated by UV radiation, emits UV radiation with a range of 320 to 440 nm, peaking at 360 nm. Because a significant amount of visible light is emitted along with the UV radiation, these bulbs are often made of a purple-red filter glass similar to that used over the high-pressure arc lamps previously described. This greatly reduces the UV radiation emitted, but still leaves what is often an excessive amount of visible blue light, considering the relatively low intensity of the UV radiation produced.

Safety Precautions for UV Lamps

As stated in the previous paragraphs, the UV lamps used for inspection utilize the long wavelength (320 to 450 nm). The use of the long wavelengths eliminates the concern for sunburn or other serious effects, such as eye irritations or skin cancer. Thus, the technician should always follow any instructions provided by the lamp manufacturer.

Testing in Subdued White Light

The improved liquid penetrant processing materials available now have made it possible to test in less than total darkness, so tests can often be carried out under far from ideal conditions. This does not mean darkened inspection booths are no longer necessary. It does mean that with bright liquid penetrant indications and adequate UV radiation intensity, medium and large indications can be detected under conditions of fairly high ambient white light. This has made some fluorescent PT possible on aircraft outdoors on the flight line. However, ambient white light requirements must be met by using black cloth or other light shields.

Although extremely dark testing areas and high UV radiation levels are not always necessary, they are, however, always helpful and do ease testing considerably. Further, in tests for the very smallest indications such as microcracks in jet turbine blades, they are necessary. In many instances, minute indications could only be detected by a dark-adapted observer working in a booth with no more than 1 fc (10 lx) of white light and then only by holding the part 2 or 3 in. (5 or 7.5 cm) from a 100 W mercury arc source, producing 150 to 180 W/m² intensities on the test object.

Liquid Penetrant Testing Equipment

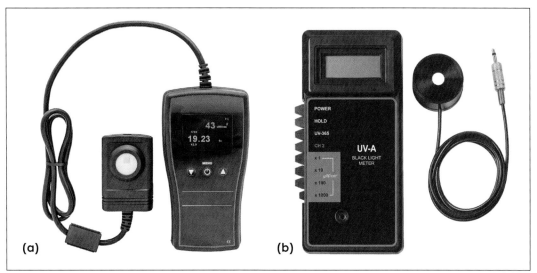

Figure 16. Photoelectric instruments for measurement of UV radiation intensity: (a) dual wavelength radiometer/photometer; and (b) UV meter.

The window lens of the visible radiation (white light) sensor must not fluoresce. Under UV-A of 3500 $\mu W/cm^2$, visible radiation of 440 fc (4.8 lx) has been reported instead of 1.2 fc (13 lx) as measured with a photometer with a nonfluorescing sensor window.

As an example, a UV radiation illumination level of approximately 800 to 1000 $\mu W/cm^2$ is sufficient to reveal most anomaly indications. This will depend, of course, on the size of the indication and type of liquid penetrant system used. There will be applications where extremely high levels of UV radiation intensity are required.

Light Meters
White light and UV radiation are measured with different sensors and in different units of measurement. A standard white light sensor measures in footcandles or lux, and the UV radiation sensor measures in microwatts per centimeter squared ($\mu W/cm^2$). Both meter and digital type instruments are available, as shown in Figure 16.

Materials For Liquid Penetrant Testing
The materials used in PT include penetrants, emulsifiers, removers or cleaners, and developers. Penetrants, emulsifiers, removers, and cleaners are furnished in liquid form. Developers are furnished in powder form. The powders may be used in the dry state, or are mixed with a suitable liquid (usually water) before use. Most penetrants, cleaners, and developers are available in pressurized spray cans, as well as in bulk. Concentrations, usage, and maintenance are in accordance with the manufacturer's directions. All bulk materials in use require a daily check for contamination.

Precleaning and Post-cleaning Materials

Special precleaning and post-cleaning materials are not required for most test objects. However, the cleaning materials must be compatible with the penetrant. Manufacturers will recommend an approved cleaner for their products. Commercially available solvent cleaners are usually satisfactory. Some procedures require sulfur and halogen-free cleaners for nickel and titanium alloys. Some test objects may have special cleaning requirements that will be specified in the company procedure.

Water-washable Penetrants

Water-washable penetrants are highly penetrating, oily liquids containing a built-in emulsifying agent that renders the oily media washable with water. The simplest to use are visible penetrants or color contrast penetrants, because no fluorescent lighting is required. They contain a dye (usually a bright red, but sometimes a special color such as blue) that can be seen under white (visible) light.

Greater visibility is obtained if fluorescent penetrants are used and viewed under UV radiation. The color of fluorescence is usually a brilliant yellow-green. For special applications, there are fluorescent penetrants that glow red or blue.

Postemulsification Penetrants

Postemulsification and solvent-removable penetrants have similar formulations to those of water-washable penetrants, except they do not contain an emulsifying agent and consequently are not soluble in water. These penetrants must be treated with a separate emulsifier before they can be removed by a water rinse or wash, or they can be removed using an approved solvent remover or cleaner.

Postemulsification penetrants are available as either visible or fluorescent penetrants. These penetrants have the advantage of eliminating some of the danger of over rinsing.

Emulsifiers

When applied to a postemulsification penetrant, emulsifiers combine with the penetrant to make the resultant mixture water-washable. The emulsifier, usually dyed orange to contrast with the penetrant, may be either lipophilic (oil-based) or hydrophilic (detergent water-based).

Lipophilic emulsifiers are usually used as contact emulsifiers, meaning that they begin emulsifying on contact with the penetrant. Emulsifiers are never applied by brushing. After the proper emulsification time, the mixture of penetrant and emulsifier can be removed with a standard water rinse. Lipophilic emulsifiers remove excess surface penetrant, allowing the remaining penetrant to be water-washable.

Hydrophilic emulsifiers (removers) can also be used as contact emulsifiers, but the emulsifier is diluted with water for tank use or sprayed as a water mix rinse. These penetrants have the advantage of eliminating some of the danger of over rinsing. Hydrophilic emulsifiers require agitation to allow fresh emulsifier to contact the surface penetrant.

Solvent Removers

Solvent removers or cleaners are used to remove excess penetrant from test surfaces. In selecting a solvent remover, only those materials approved by the penetrant manufacturer and penetrant procedure can be used. These solvents are available in aerosol cans or bulk.

Dry Developer

Dry developer is a fluffy powder that is applied to dry test surfaces (after the removal of excess penetrant) for the purpose of absorbing penetrant from discontinuities and enhancing the resultant indications. Of the different developers available, dry developer is the most adaptable to rough surfaces and automatic processing. It is also the easiest to remove. Dry developer is less sensitive than the other types of developers, but provides better resolution and is better suited for certain applications, such as fine threads or rough castings. Dry developer is also much easier to remove after the test.

Nonaqueous Wet Developer

Nonaqueous wet developer is a suspension of developer particles in a rapid-drying solvent. It is most often used with solvent-removable processing. Like dry developer, nonaqueous wet developer is applied only to dry surfaces. Of all the developers, nonaqueous wet developer is the most sensitive in detecting fine discontinuities. The evaporation of the solvent carrier helps to draw the penetrant from discontinuities.

Water-based Developers

Water-based developers function similarly to dry developer, except they are applied before drying the test object. Two types of developer are available. In the water-suspendable developer, particles are held in water and require continuous agitation to keep the particles in suspension. In water-soluble developer, powder is dissolved in water, forming a solution; once mixed they remain mixed. Water-based developer requires a periodic check (usually weekly) with a hydrometer for concentration.

Special Purpose Penetrant Materials

In addition to the conventional penetrants, emulsifiers, removers, and developers used in PT, there are low sulfur and low chlorine materials for testing nickel alloys, certain stainless steels, and titanium. Special purpose inert materials are available for testing objects that come in contact with liquid oxygen, rubber, or plastic.

Food-compatible materials are also available. There are high temperature penetrants for testing hot welds, and special penetrants for testing at low temperatures. There are supersensitive penetrants for detecting extremely fine discontinuities, and penetrants that provide sufficient contrast and sensitivity without a developer.

There are low energy emulsifiers and inhibited solvent removers to slow emulsification and the removal of excess penetrant. There are also wax and plastic film developers that absorb and fix penetrant indications to provide record of the test.

The selection and usage of these materials is largely dependent on the particular process used and the controlling specifications or standards. All special purpose materials require specific approval for industrial applications.

Additional Precautions

The PT technician must be sure that the test object is not damaged or overheated during the test. If any test object contains plastic, rubber, or other nonmetallic components, the technician must check to see if solvent will harm that test object.

There are also ecological requirements for disposal of penetrant residue and rinse water. The solvents and aerosol cans are flammable, and normal precautions for flammable liquids apply.

The materials used in testing are manufactured with consideration to possible health hazards. These products are qualified as safe for use by humans in industrial operations. Liquid penetrant materials and solvents require standard safety procedures, as detailed by their manufacturers. The technician should always follow any instructions provided by the material manufacturer.

LEVEL ►||

Selection of Liquid Penetrant Testing Method

Selection

Selection of penetrant material combinations of penetrant, emulsifiers, penetrant removal methods, and forms of developers must be made in accordance with the penetrant material manufacturer's recommendations and instructions or an applicable qualified products list. The required liquid penetrant type and method are sometimes decided on by the contractor, customer, Level III technician, or design engineer. In some cases, managers have the option of using different combinations of penetrants, emulsifier systems, penetrant removal methods, or developers.

The selection of a suitably efficient, cost-effective penetrant type and process for a particular application depends on several factors:

- customer requirements
- specification requirements
- sensitivity requirements
- size and number of test objects (see Figure 1)
- surface condition of the test objects
- configuration of the test objects
- cost of equipment and materials
- availability of water, electricity, compressed air, and a suitable testing area
- engineering design requirements

Figure 1. Multiple small parts being rinsed in basket after dwell time and processed at one time.

Advantages and Disadvantages

Each technique has its own advantages and disadvantages that can be utilized to address a potential application. As follows is a list of the advantages and disadvantages of each penetrant type, method, sensitivity, and developer. Table 1 shows a comparison between the different types and methods used in PT.

Table 1. Advantages and disadvantages of the different types and methods of PT

Technique	Advantages	Disadvantages
Type I	Easier to see indications	Dark area and special lighting required
Type II	No special lighting required	Less sensitive indications
Type III	Flexible implementation	Decreased contrast
Method A	Fast	Danger of over washing
Methods B and D	Can achieve sensitivity Level 4 Less danger of over washing	Extra step and material required Slower
Method C	Portable and no power required	Slower

Penetrant Types

The advantages and disadvantage of the different penetrant types are as follows.

Type I Fluorescent
- Advantages: greatest contrast of the inspection material to the part.
- Disadvantages: inspections need to be incorporated in darkened areas with specialized UV lighting equipment. Also required is the utilization of appropriate personnel protection equipment such as UV-A radiation glasses to protect the technician's eyes.

Type II Visible
- Advantages: no special lighting is required.
- Disadvantages: does not have the sensitivity or contrast that Type I achieves.

Type III Dual-mode (Visible and Fluorescent)
- Advantages: can be implemented in conditions ideal for either Type I or Type II inspections.
- Disadvantages: does not provide the same amount of contrast for similar Type I or Type II inspections.

Penetrant Method

The advantages and disadvantages of each method are as follows.

Method A Water-washable
- Advantages: typically used for applications like large castings, forgings, or small machined test objects. It comprises a faster, chemically safer, more ecofriendly, and inexpensive process. It also does not make use of an additional chemical (emulsifier) and can be utilized as a portable method.
- Disadvantages: usually a less sensitive method and there is a danger of overwashing the penetrant out of potential cracks.

Method B Postemulsifiable (Lipophilic)
- Advantages: because of the necessary monitoring, there is a reduced risk of washing penetrant out of shallow discontinuities. In addition, if the technician deems there was inadequate emulsification to the test part, the part can be reprocessed to obtain additional emulsification.
- Disadvantages: the high cost of the emulsifier and equipment, and the additional process control checks. Also of note is the extra step of emulsification and thus the additional material required. Additionally, Method B is not permitted by some industry specifications for aircraft engine parts.

Method C Solvent-removable
- Advantages: a portable technique that takes very little setup time to incorporate an inspection. Additionally, if a Type II solvent-removable inspection is chosen, no power or water supply is needed. This method also does not require frequent process controls that other methods may require.
- Disadvantages: the manual wipe is a slower, less sensitive method. It is also difficult to utilize on large inspection surfaces or batches of parts.

Method D Postemulsifiable (Hydrophilic)
- Advantages: for large test objects, equipment that sprays water and Method D (hydrophilic) emulsifier mix may be more efficient than other methods. There is a reduced risk of washing penetrant out of shallow discontinuities and it is one of the more sensitive methods.
- Disadvantages: the cost of the emulsifier and equipment, and the need for precision timing for the emulsification. Additionally, more process controls and maintenance checks are needed to ensure the system is working properly.

Figure 2 shows an additive manufactured part, a laser powder bed fusion Ti-6Al-4V, post-machined and designed for the *V-22* Naval aircraft. This part was inspected with Type I, Method D, sensitivity Level 4. The inspection allowed for the detection and characterization of a 1.25 in. (3.18 cm) crack in the base of the fitting. With the implementation of the fluorescent PT, the contrast between the part and crack indication is significant.

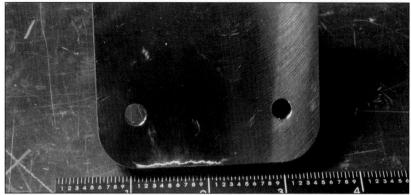

Figure 2. Additive manufactured part inspected with Type I, Method D, Sensitivity 4. (Courtesy of the NDI branch, Materials Division, Naval Air Warfare Center Aircraft Division at the Patuxent River Naval Air Station, 2016).

Figure 3. Sensitivity levels of two different penetrants: postemulsifiable penetrant at Level 3 sensitivity (left) and water-washable penetrant at Level 2 sensitivity.

Penetrant Sensitivity

With any penetrant type, process, method, or sensitivity (Figure 3), overcleaning can occur if technicians fail to follow the appropriate work practices. The advantages and disadvantages of lower and higher sensitivity penetrants are as follows.

Lower Sensitivity Penetrants (1/2, 1)
- Advantages: there is a decrease in background indications.
- Disadvantages: lower sensitivity inspections (cannot reliably detect smaller discontinuities).

Higher Sensitivity Penetrants (3, 4)
- Advantages: higher sensitivity and can effectively detect smaller discontinuities.
- Disadvantages: increased background and nonrelevant indications. This can decrease the contrast of indications.

Developer

Choice of developer for all the methods is regulated by specifications and the manufacturer's recommendation. The selection and usage of these materials is largely dependent on the particular process used, controlling specifications or standards, and company policy regarding recording of test indications. The advantages and disadvantages of different developers are as follows.

Form A Dry Developer
- Advantages: best for rough surfaces, such as castings or test objects with fine threads or corners such as keyways. It is also easy to handle, transport, and apply. Dry developer leaves no film; thus, no special cleaning is required for subsequent processing operations and the process is generally inexpensive. Dry developer can be applied by blowing, immersion, pouring, fog chamber, or spraying.
- Disadvantages: test objects must be completely dry before application. Excessive drying time will reduce sensitivity. Dry developer is the least sensitive type of developer and should not be used for visible (Type II) penetrant. The use of dry developer may require the use of respirators.

Form B Water-soluble
- Advantages: may be applied by dipping, spraying, or flowing. The part during application can be wet or dry. There is no need for agitation to keep developer particles in a suspension and the particles are easy to remove in post-cleaning of the test part.
- Disadvantages: indications can be blurred or indistinct if not dried quickly. Form B requires daily maintenance checks when used in tanks, and is typically not used with water-washable and/or visible dye penetrants.

Form C Water-suspendable
- Advantages: may be applied by immersion, spraying, or flowing. The part during application can be wet or dry. The developer particles are insoluble in water and when dried are highly absorptive.
- Disadvantages: requires frequent stirring or agitation to ensure developer particles do not settle. Additionally, the bath requires daily maintenance checks when used in tanks. Use of this developer may produce streaks or runs.

Forms D (Fluorescent) and E (Visible) Nonaqueous
- Advantages: capable of being the most sensitive forms of developer. They are portable as used in an aerosol can. The advantages of use through an aerosol can are that there is minimal risk of contamination, and the thickness and area of application can be easily controlled by the operator. The developer dries easily, and therefore has no need for a drying oven.

- Disadvantages: typically only applied to a test surface with a spray gun or aerosol can. The test object must be completely dry before application. Agitation (shaking) of the container prior to application is required to unsettle developer particles. The use of cans yields small spray area coverage and, as such, coating a large surface would be very time consuming.

Form F Special Application
- Advantages: used for unique inspections that are not appropriate for the other methods due to extreme environmental conditions or unique needs for reporting results. Some special applications provide fixed penetrant indications to provide permanent records of discontinuities.
- Disadvantages: may lack the sensitivity and clarity of more conventional developers.

Interpretation and Evaluation of Indications

Discontinuity Categories

Specific discontinuities are divided into three general categories: inherent, processing, and service. These categories are further classified as to the stage of manufacturing, material (ferrous or nonferrous), and manufacturing process. No matter what type of discontinuity may be present in the test object, only those open to the surface may be detected with PT.

Inherent Discontinuities

In metals, inherent discontinuities are those that are related to the melting and original solidification of the molten metal, ingot, or casting (See Figures 1 and 2).

- Ingot discontinuities are those related to the melting, pouring, and solidification of the original ingot, and include shrinkage, slag, porosity, cracks, and nonmetallic inclusions.

Figure 1. Blowhole indications in a nuclear power plant casting component.

Figure 2. Liquid penetrant indications of seams.

- Inherent cast discontinuities are those related to the melting, casting, and solidification of the cast test object. They include those discontinuities introduced by casting variables, such as inadequate feeding, gating, excessively high pouring temperature, entrapped gases, shrinkage, hot tears, inclusions, cracks, and blowholes.

Processing Discontinuities

Processing discontinuities are related to the various manufacturing processes, such as forging, machining, forming, extruding, rolling, welding, heat treating (Figures 3 to 5), and plating. These discontinuities may be caused by the processing technique, or they may be inherent discontinuities that have been exposed or changed (in shape or direction) by the processing technique.

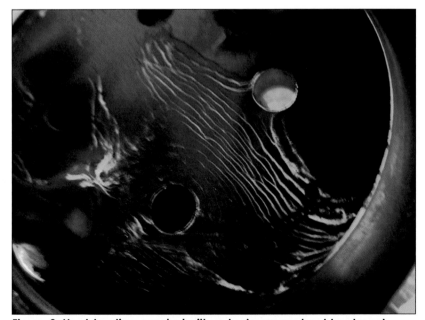

Figure 3. Heat-treating cracks built up by improper heat treatment.

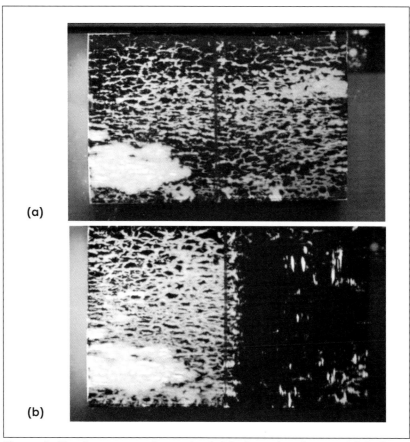

Figure 4. Aluminum quench cracks: (a) before cleaning; and (b) after mechanical cleaning with a plastic pad.

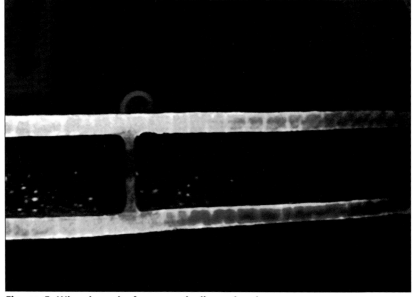

Figure 5. Wheel marks from a grinding wheel.

Figure 6. Fatigue crack propagating from bolt hole down the side of an aircraft wheel.

Service Discontinuities

Service discontinuities (Figures 6 to 10) are related to service conditions such as cycles of loading, stress corrosion, fatigue, and wear.

Indications

Knowledge of the different types of indications a penetrant inspector might encounter during the evaluation process will offer a more complete inspection. Knowing how these indications occur and what they might look like is very important.

Forming of Discontinuity Indications

Discontinuities that have an opening to the surface can be detected using PT. After the proper precleaning, drying, and application of penetrant for the proper dwell time (typically

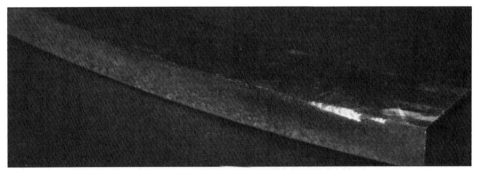

Figure 7. Fluorescent indications of tears in stretch formed aluminum angle.

Figure 8. Fluorescent penetrant indication of a fatigue crack.

Interpretation and Evaluation of Indications

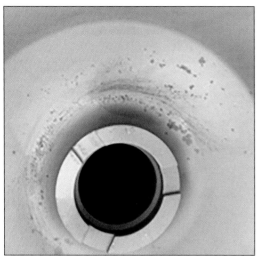

Figure 9. Corrosion of aluminum in the base of a propeller assembly.

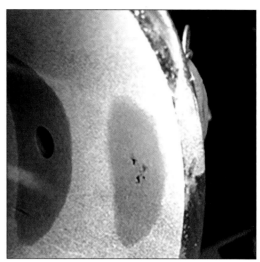

Figure 10. Persistent bleedout of a deep corrosion indication.

approximately 10 min), capillary action forces the penetrant into the discontinuity. Then, the excess surface penetrant is removed either by wiping or with a water spray rinse, with the aid of an emulsifier for postemulsifiable penetrants.

Time for Indications to Appear

After removal of excess surface penetrant and application of developer, penetrant will migrate to the surface aided by capillary action and the blotting action or the dilution action of the developer. Deep indications will begin to appear first since indications with a large reservoir of penetrant will bleed out faster than small indications. The typical time to allow indications to form before measuring and evaluation is specified in the written procedure. Standard times for evaluation are important so that different technicians and testing facilities will get standard results.

Persistence of Indications

One technique to verify that a questionable indication is relevant is to wipe the penetrant away with a swab dampened with solvent. The persistence of a relevant indication will cause it to bleed out again. Caution must be used to ensure that the swab is only dampened and not dripping with solvent.

Effects of Temperature

The standard temperature range for PT is 50 to 100 °F (10 to 38 °C). Colder temperatures will increase the viscosity of the penetrant and slow the capillary action. Some specifications allow temperatures of 40 to 150 °F (4.4 to 65.5 °C), but require doubling the dwell time for temperatures below 50 °F (10 °C).

Temperatures higher than approved may cause the penetrant to dry and reduce sensitivity. Some specifications allow special, high-temperature materials that are available for temperatures up to 350 °F

(177 °C). High-temperature materials require special procedures, special training, and certification of technicians.

Lighting

The standard lighting for viewing and evaluating visible dye penetrant indications is a minimum of 100 fc (1000 lx) of white light at the test surface. For fluorescent penetrant indications, the standard lighting is a minimum of 1000 μW/cm^2 UV radiation at the test surface, and a darkened test area of less than 2 fc (20 lx). The standard lighting in a fluorescent rinse or penetrant removal area is less than 10 fc (100 lx) and greater than 100 μW/cm^2.

Effects of Metal-Smearing Operations

Operations such as power wire brushing or sand blasting can smear metal and close the surface opening of discontinuities (Figure 11). If these operations have been performed on a test object, the surface must be etched to remove the smeared metal. Samples of the visual appearance of smeared metal should be available or used in technician training.

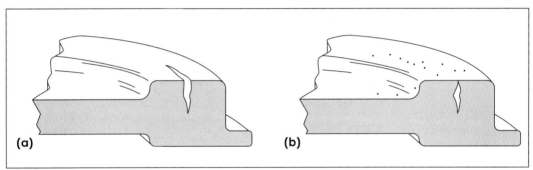

Figure 11. The effect of mechanical cleaning operations on discontinuities: (a) before mechanical cleaning; (b) after mechanical cleaning.

Sequence

In-process tests are sometimes used for the root layer of welds, or before final machining. Final PT is performed on a test object in the final machined and heat-treated condition after proper precleaning.

Test Object Preparation

The test object must be processed through preparation, such as etching, precleaning, and drying, as efficiently as possible. All parts should be properly cleaned and dried prior to penetrant processing.

Factors Affecting Indications

Factors that affect indications include the penetrant used, prior processing, and surface condition of the part.

Penetrant Used

The choice of penetrant depends on the sensitivity requirements, location, availability of UV radiation, water, power, and test requirements.

Fluorescent indications are easier to see, and the postemulsified method is the most sensitive for small, shallow indications.

Small shallow indications can be viewed under a 10× magnifying glass or magnifier on a support stand and should be included in the inspection area. For more critical viewing, the value of using a microscope (with 20 to 100× enlargements) for the viewing of micro-indications as well as for determining the extent and nature of macro-indications (with 5 to 20× enlargements) cannot be stressed enough (Figure 12).

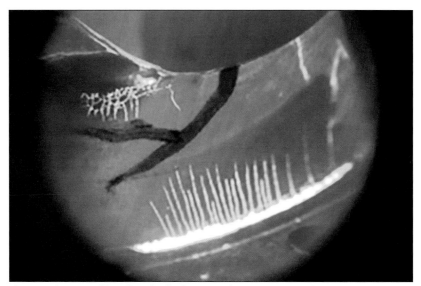

Figure 12. Magnification (5×) of cracks in a chrome plate.

Prior Processing

The precleaning technique may depend on prior processing of the test object. If a test object or weld has never been exposed to machine oils or lubricants, some specifications will allow reduction of penetrant dwell time to as little as 5 min. Ultrasonic testing (UT) using a couplant should only be performed after final PT because the couplant may hinder penetrant penetration into discontinuities. As mentioned in Chapter 3, visible dye penetrant should never be used before a fluorescent penetrant test because the red penetrant may still be present in discontinuities and could interfere with indication luminosity.

Surface Condition

The surface condition of the test object can interfere with testing. A surface opening may be closed by lubricants, polishing compounds, dirt, scale or other contaminants that are forced into the cracks or holes.

Rough or porous surfaces may retain penetrant, resulting in nonrelevant indications. These same surfaces can retain other oils or greases that can fluoresce and cause confusing indications.

Moisture or water within a discontinuity cavity on the part can prevent penetrant from entering the discontinuity. Other deposits may dilute the penetrant and reduce the effectiveness by either destroying the fluorescence or dye color.

Crack Indications
Cracks can occur at any stage of manufacture or service. An indication from a crack will be linear (three times as long as wide) and slightly irregular. Test objects that have cracks are usually rejected.

Solidification Cracks
Ingot cracks occur during metal solidification, and may change to a straight line during processing. They appear as straight line indications in the direction of rolling or working.

Processing Cracks
A typical processing crack can occur during welding, straightening, bending, or from internal stresses at any point during processing, including heat treatment or plating. These cracks will also be linear and slightly irregular, and test objects that have processing cracks are routinely rejected.

Service Cracks
Cracks (shown in Figures 13 to 16) can occur at any time during service for a multitude of reasons. Usually they are caused by overloading, fatigue or, in some cases, corrosion such as stress corrosion cracking. Test objects that have in-service cracks are typically rejected.

Figure 13. Fatigue crack running through gear tooth propagating through an aluminum case.

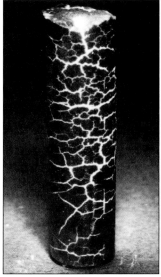

Figure 14. Fluorescent indications of cracks produced in a hard fired unglazed ceramic rod.

Interpretation and Evaluation of Indications

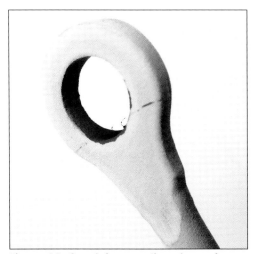

Figure 15. Crack in a casting shown by visible dye liquid penetrant testing.

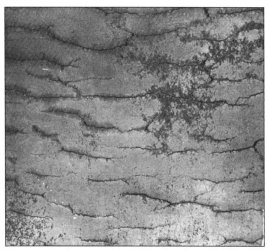

Figure 16. Dye penetrant indications (without developer) of stress corrosion cracking of a metallic sheet.

Porosity Indications

Liquid penetrant indications from porosity are rounded (less than three times as long as wide) and can occur in the ingot, casting, or weld. Typically, no porosity is present or allowed in forged, rolled, or extruded materials. Some small porosity is acceptable to some specifications, especially for castings and welds.

Indications from Specific Material Forms

Not only are seeing and defining the type of indication very important, but so is knowing how the indication occurred in the manufacturing process. This knowledge will help determine if an indication is a discontinuity or just inherent to the type of manufacturing.

Forgings

Laps (Figure 17) are caused when metal is folded over and not fused either at transition areas or from improper trimming at flash line areas. Laps will be open to the surface and linear (three times as long as wide), with some shape but no particular orientation to the direction of forging.

Bursts are caused by forging at an improper temperature. Bursts can be at the surface if the surface temperature is incorrect, or internal if

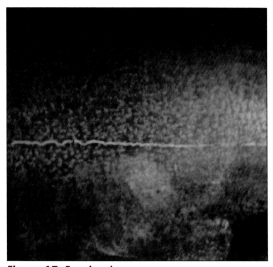

Figure 17. Forging lap.

63

the internal temperature is incorrect. Bursts are linear and branch or chevron-shaped.

Most critical forgings are tested for internal discontinuities with UT. PT is typically used for nonmagnetic materials. Magnetic particle testing (MT) is often used for ferromagnetic forgings to detect surface indications.

Castings

Porosity or trapped gas pockets that could not escape at the risers can form either at the surface or within the interior of the test object. Surface connected porosity (Figures 18 to 20) will produce rounded indications.

Shrinkage (Figure 21) is caused by stresses from unequal cooling. Surface shrinkage can be detected visually or by PT of the surface. Shrinkage is normally internal and detected by radiographic testing (RT), unless shrinkage is machined open to the surface.

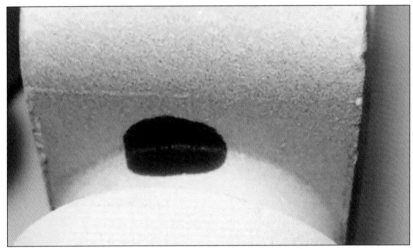

Figure 18. Visible dye indication of porosity in a casting.

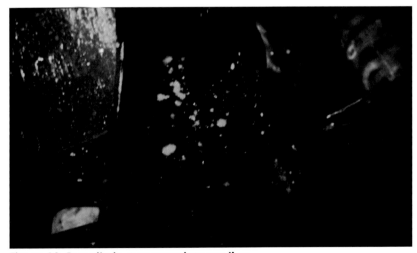

Figure 19. Porosity in a magnesium casting.

Interpretation and Evaluation of Indications

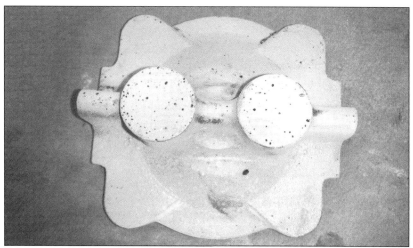

Figure 20. Visible dye penetrant indications of surface porosity in a casting.

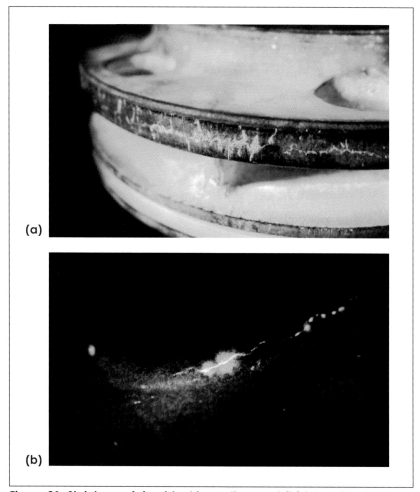

Figure 21. Shrinkage: (a) evident in casting; and (b) intermittent shrinkage in a magnesium cast case.

Hot tears (Figure 22) are caused by unequal cooling between thick and thinner sections of castings. Indications are linear and crack-like, but usually wider than a crack. Hot tears are typically open to the surface, but may be inaccessible for PT. RT can detect hot tears.

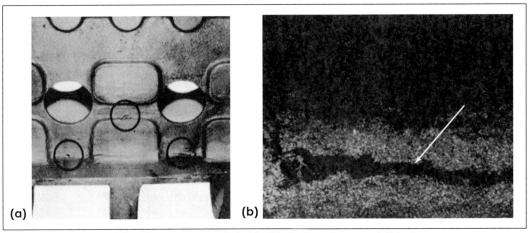

Figure 22. Hot tears (visual appearance): (a) in fillet of casting; and (b) close-up.

Cold shuts (Figure 23) are caused by metal solidification before the metal melts together. This is caused by splashes or streams of molten metal meeting and solidifying, instead of flowing together. Cold shuts may be open to the surface. They can be detected by RT.

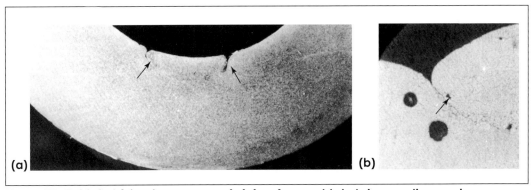

Figure 23. Cold shut (visual appearance): (a) surface cold shuts in a casting; and (b) micrograph of the cross section of a cold shut.

Critical castings are often tested with RT for internal discontinuities because the large material grain sizes make UT difficult. PT is used to detect surface discontinuities in nonferromagnetic castings, and MT is used for ferromagnetic castings.

Plate

Laminations (Figure 24) are the most common form of discontinuities in plate. They are caused by relatively large discontinuities, such as

Interpretation and Evaluation of Indications

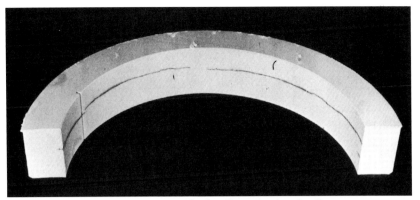

Figure 24. Visible dye penetrant indication of a lamination.

blowholes, slag, or inclusions, which are rolled flat and parallel to the surface and are open to the surface at the end or edge of the plate. UT is used to detect discontinuities in critical plate. Some laminations are acceptable if they are outside a weld area and meet certain spacing and size criteria.

Welds

Porosity (Figures 25 and 26) is a discontinuity in welds that can be open to the surface or subsurface. Surface porosity will appear as a round indication. Porosity can be caused by improper welding technique or moisture, oil, or foreign material in or near the weld metal.

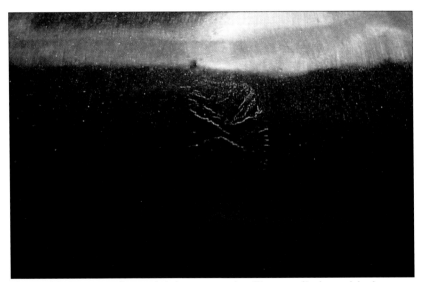

Figure 25. Large surface shrinkage crack with porosity in welded tube stock.

Crater cracks (Figures 27 and 28) or crater pits are found where the welder stopped a length of weld and used an improper technique for stopping. Crater pits may go to the root of the weld, and both conditions will result in a relevant liquid penetrant indication.

Liquid Penetrant Testing Classroom Training Book | CHAPTER 6

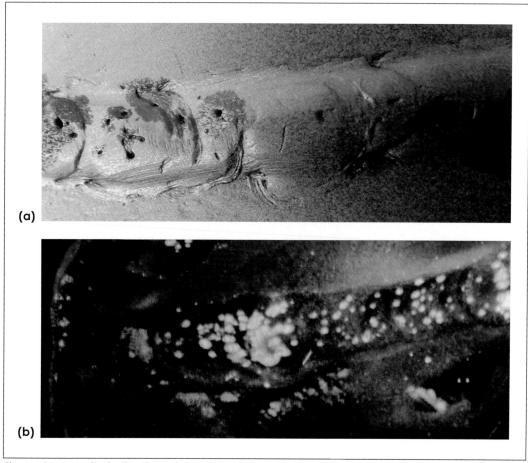

Figure 26. Porosity indications: (a) visible dye; and (b) fluorescent penetrant.

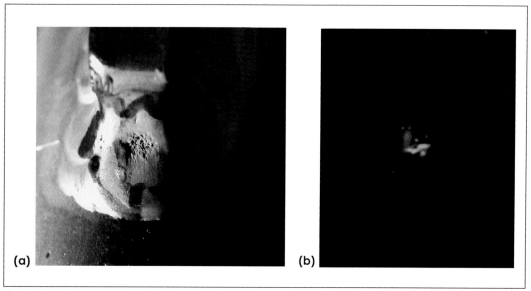

Figure 27. Crater crack in weld: (a) visual appearance; (b) fluorescent penetrant indication.

Interpretation and Evaluation of Indications

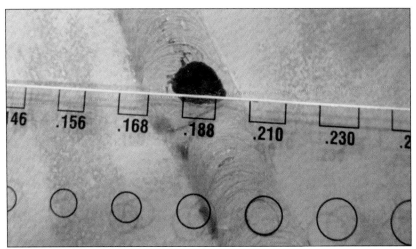

Figure 28. Visible dye liquid penetrant indication from a crater crack.

Weld fusion lines can be difficult to clean properly if acceptable weld undercut is present. Weld undercut occurs where the weld has not completely refilled the edge of the melted base material. This area is where fusion line cracks can also occur, so care must be used when evaluating this area. In some cases, additional grinding must be done on visually acceptable undercut to allow proper penetrant testing.

Transverse weld cracks can also occur usually from a filler material problem or hydrogen embrittlement.

Extrusions
Discontinuities in extrusions and in rolled shapes are usually longitudinally aligned, parallel to the direction of working. Acceptance criteria for forgings are used.

Evaluation of Indications
To evaluate an indication means to decide if the indication is acceptable, requires rework, or causes the test object to be rejected.

False Indications
The most common source of false indications (Figures 29 and 30) is poor washing of water-washable and postemulsified penetrants. When using fluorescent penetrant, UV radiation during the washing process is important. The technician must determine whether a good rinse is obtained or whether patches of fluorescence remain on the test object. Adequate lighting at a fluorescent rinse or cleaning station is usually considered less than 10 fc (100 lx) visible light and 100 $\mu W/cm^2$ UV radiation at the test surface.

Solvent removal is slower and more thorough. To guard against confusion resulting from fluorescent or color spots other than relevant indications, care is taken so that no outside contamination occurs. Typical sources of contamination include the following:
- penetrant on the hands of the technician
- contamination of wet or dry developer

- penetrant rubbing off of one test object to a clean portion of the surface of another test object
- penetrant spots on the testing table

To avoid contamination, process tanks and testing areas are kept clean, only lint-free wiping cloths are used, and test objects are kept free of fingerprints and tool marks.

Relevant Indications

Relevant indications (Figure 31) are those caused by a material discontinuity. The interpretation of an indication as relevant is a matter of observing the indication, eliminating the possibility of it being a false

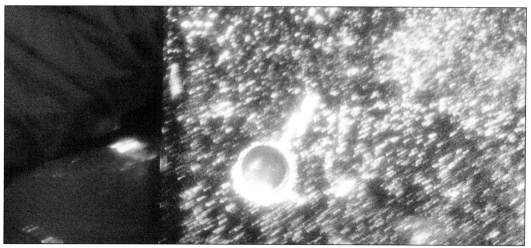

Figure 29. False indications in porous aluminum forging.

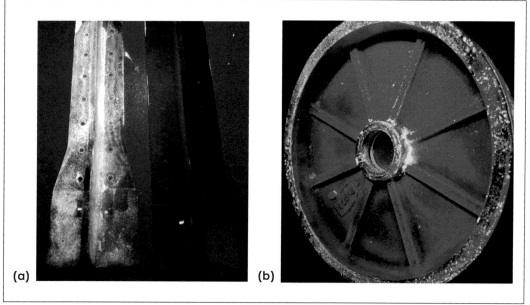

Figure 30. False indications: (a) from inadequate penetrant removal; (b) from developer contamination.

Interpretation and Evaluation of Indications

Figure 31. Relevant fluorescent indication from a fatigue crack.

indication, and then further determining that it is not a nonrelevant indication. Any relevant indication immediately becomes subject to interpretation of its cause (type of discontinuity).

Nonrelevant Indications

Nonrelevant indications (Figures 32 and 33) are caused by features in the test object that are there by design, but are in no way relevant discontinuities. Most nonrelevant indications are easy to recognize because they are related directly to some feature of the assembly that accounts for their presence.

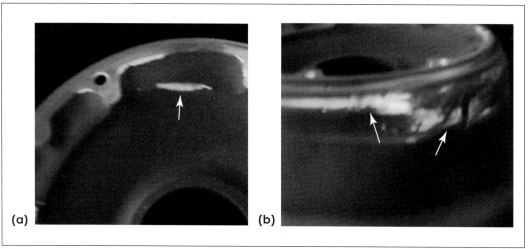

Figure 32. Nonrelevant fluorescent penetrant indications on aircraft wheel: (a) inner surface; and (b) outer surface. Visual testing shows the surface condition.

Figure 33. Nonrelevant indications from a relevant linear indication from an in-service fatigue crack (left) compared with threads and number stamping.

Nonrelevant indications include those that appear on test objects that are press fitted, riveted, or spot welded together, and those caused by surface roughness.

Any nonrelevant indication that interferes with or could mask a relevant indication must be further evaluated and sometimes retested. If a test object is too rough to perform a proper liquid penetrant test, or has a condition like weld undercut that has been accepted visually, the test object or weld may have to be returned for further preparation for PT.

7
Liquid Penetrant Process Control

Quality Control of Test Materials
The reliability of any penetrant test is determined in large part by the condition of the materials used. Even the best procedures are ineffective if test materials are faulty. To ensure the satisfactory condition of penetrant test materials, various quality control procedures are used.

This chapter discusses the in-service checks used to test penetrant materials held in open tanks and subject to contamination or evaporation. Each of the quality control procedures mentioned here is based on the assumption that care, handling, and use of the materials are strictly in accordance with manufacturers' recommendations.

Many additional quality control procedures, such as the determination of sulfur and chlorine content, liquid oxygen compatibility, temperature stability, water tolerance, viscosity, flashpoint, toxicity, and developer precipitation rates, are described in the various controlling specifications and standards. These procedures, however, are primarily of interest to the manufacturer or laboratory technician rather than the individual performing or monitoring penetrant tests. Quality control procedures are also available from manufacturers and various technical societies such as ASTM.

Test Material Control Samples
Many of the tests used in the control of penetrant materials are comparison tests in which used materials are compared with new materials. For this reason, control samples are taken at the time the materials are received from the supplier. These samples are kept in sealed containers and stored where they are not subject to deterioration from heat, light, or evaporation. The procedures for taking the samples and their renewal are typically detailed in company procedures, controlling specifications, and standards.

Reference Blocks
The use of reference blocks, plates, or panels is often specified in procedures for quality control of liquid penetrant materials. The materials used in the manufacture of reference blocks include aluminum, steel, nickel, glass, and ceramic. Some of the blocks are designed primarily for checking penetrant or system sensitivity and performing comparison tests, whereas others are designed specifically for testing penetrant or emulsifier washability. All, however, are prepared to rigid

specifications as required for specific tests, as described in the following sections.

System Monitor Panels

Liquid penetrant system monitor panels, as shown in Figure 1, are commercially manufactured and can be used to detect major changes both in visible and fluorescent systems, and in water-washable and postemulsified systems. The liquid penetrant system monitor panel is normally processed at the beginning of each shift to verify system performance.

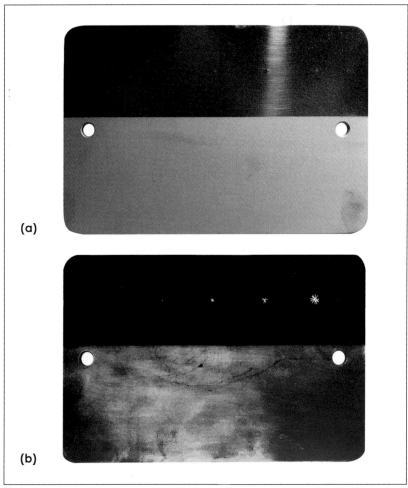

Figure 1. A liquid penetrant system monitor panel includes five crack centers of different sizes for evaluating sensitivity and includes a grit-blasted section for judging wash characteristics of the liquid penetrant system: (a) under white light; and (b) under UV radiation.

The panel is stainless steel, approximately 0.1 in. (0.25 cm) thick and 4 × 6 in. (10 × 15 cm). A strip of chrome plating runs the length of the plate, and the other half of the same side of the plate is an oxide grit-blasted surface to monitor background color or fluorescence. The chrome strip has five variable size crack centers induced by a hardness punch from the reverse side. The largest crack pattern is readily visible with low sensitivity penetrant materials, and the smallest is difficult to observe even with high sensitivity penetrant materials.

Aluminum Reference Blocks

Aluminum reference blocks measure approximately 2 × 3 in. (5 × 7.5 cm), and are cut from 0.3 in. (0.8 cm) thick bare 2024-T3 aluminum alloy plate, with the 3 in. (7.5 cm) dimension in the direction of rolling. The dimensions are for guidance only and are not critical.

The blocks are heated and water quenched to produce thermal cracks. This is accomplished by supporting the block in a frame and heating it nonuniformly with the flame of a gas burner or torch in the center on the underside of the block. The flame remains centered and does not move in any direction during the heating process.

A 950 to 980 °F (510 to 527 °C) temperature-indicating medium—typically in the form of lacquer, pellets, or sticks—is applied to an area the size of a penny on the top side and directly in the center of the block. The heat of the torch or burner is adjusted so that the block is heated approximately 4 min before the temperature-indicating medium melts, after which the block is immediately quenched in cold water, as shown in Figure 2.

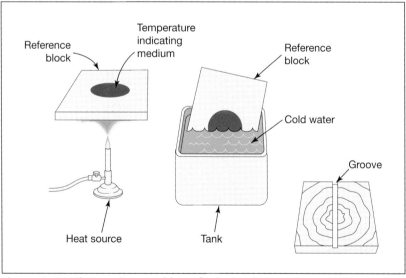

Figure 2. Heating and quenching of an aluminum reference block.

A groove approximately 0.06 × 0.06 in. (0.15 × 0.15 cm) deep is cut in the 2 in. (5 cm) direction across the center of the heat-affected zone. This forms two test areas, and permits the side-by-side application and comparison of two penetrants without cross-contamination. This type of block is widely used for comparing the performance of penetrants in field conditions.

Preparation for Use
Before use, aluminum reference blocks are scrubbed with a bristle brush and liquid solvent, followed by degreasing and drying.

Preparation for Reuse
After a reference block has been used, it is cleaned before reuse. The block is heated slowly with a gas burner to 800 °F (426 °C), as determined by a temperature-indicating medium, after which the block is quenched in cold water. It is then heated to approximately 160 °F (71 °C) for 30 min to drive off any moisture in the cracks, and is then allowed to cool to room temperature.

Ceramic Reference Blocks
Ceramic reference blocks are flat, circular disks of unglazed ceramic that, although quite solid and impervious to liquids, have micropit surfaces that entrap liquid penetrants. The micropit structure provides a range of pore sizes, and a performance comparison can be made of two or more penetrants merely by noting the number or distribution of porosity indications and their intensity in a side-by-side comparison test. Indications appear as a large number of microscopic specks of fluorescence or color, the number increasing as the sensitivity of the penetrant increases.

Preparation for Use
The ceramic reference block is removed from its container (a shallow jar containing alcohol) using a pair of tweezers to avoid touching with the fingers. The block is then wiped with a soft absorbent tissue, and allowed to dry for 5 min.

Usage
Using a small applicator, a drop of each penetrant to be tested is applied to the flat surface of the reference block. Immediately following application, the penetrants are blotted with a piece of soft tissue by pressing the tissue against the block using a flat object. The tissue and cover also hold the penetrants in proper contact with the reference block and prevent excessive bleeding and possible cross-contamination.

Following the required dwell period (usually 10 min), the penetrants are processed in accordance with the penetrant manufacturer's recommendations. A developer, however, is not used. In making visual comparisons, both the number of indications observed and the intensity of indications are noted.

Cleaning
After use, the ceramic block is placed on its edge in alcohol and soaked for several hours or until the penetrant entrapments diffuse into the alcohol. The block is then returned to its storage container, covered with fresh alcohol and sealed.

Anodized and Plated Test Panels
Stress cracked anodized aluminum and chrome-plated nickel test panels are frequently used for comparing penetrant sensitivity and washability. The panels are classified according to the size cracks they contain. The

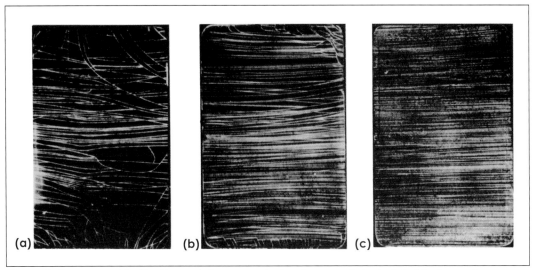

Figure 3. Fluorescent indications of stress cracks in chrome-plated nickel test panels: (a) coarse cracks; (b) medium cracks; and (c) fine cracks.

grades are coarse, medium, and fine, providing low, medium, and high sensitivity levels (see Figure 3).

Preparation for Use
Panel surfaces are scrubbed using moderate pressure and a cloth dampened with emulsifier or concentrated soap solution, followed with a thorough rinse using a water spray. After rinsing, the panel is oven dried at an approved temperature for 3 to 5 min to drive off moisture remaining in cracks. The panel is then dipped in acetone and agitated for approximately 1 min, removed, and again oven dried at an approved temperature for approximately 5 min.

Usage
A line is usually drawn along the centerline (length) of the panel using a wax pencil or narrow vinyl tape. This forms two test areas and permits the side-by-side application and comparison of penetrant materials without cross-contamination.

Twin Nickel-chromium Sensitivity Panels

A set of two nickel-chromium (NiCr) panels, each measuring 3.9 × 1.4 in. (10 × 3.5 cm) is sheared from the same stock. This allows for matching crack patterns. With a matched set of panels, a simultaneous comparison of two penetrant materials can be achieved. The panels come in sets having crack depths of 0.4, 0.8, 1.2, and 2 in. (10, 20, 30, and 50 μm).

Stainless Steel Test Panels

Stainless steel test panels or plates are used in testing the washability of fluorescent penetrants and visible dye penetrants. The panels are prepared from annealed type 301 or 302 stainless steel and measure 2 × 4 in. (5 × 10 cm) or larger. Each is sandblasted on one side with 80 mesh average size grit, at 60 psi (414 kPa), with the gun held approximately

18 in. (46 cm) from the surface. Sandblasting continues until a uniform matte surface is obtained. It is the sandblasted surface that is used in performing tests.

Preparation for Use
Before use, the panels are cleaned by degreasing, heated to 160 °F (71 °C), and then allowed to cool to room temperature in a dry area.

Low Cycle Fatigue Blocks
Titanium or nickel-chromium-iron (NiCrFe) plates are commonly used to manufacture standards with low cycle fatigue (LCF) block cracks in various size ranges. The cracks are started from electrical discharge machined notches or spot welds, which are later ground away after the starter cracks are grown. Tensile stressing or reverse bending of the plates achieves additional crack length extension. Titanium or nickel-chromium-iron plates are commonly sold in a set of three plates, with a total of 18 possible cracks in the set.

LCF blocks are used like other known discontinuity standards, except that the total number of detected cracks per inspection of the plate set is recorded and monitored in a running summary per procedure supplied with the plates. When fewer cracks are detected, the technician is warned that something has shifted in the process capability, or that the cracks have been improperly cleaned.

LCF blocks are cleaned and stored in acetone or equivalent solvent, or in a foam form made for them, as long as they are cleaned with alcohol or acetone before and after each use. LCF blocks can also be returned to the manufacturer for extensive cleaning, if necessary.

Liquid Penetrant Material Tests
The in-service quality of the materials used in PT is determined by a check of sensitivity, water content, contamination, and washability. The tendency toward fading of the penetrant dyes is also checked by a simple comparison test. The tests described as follows are typical of those performed on used or questionable penetrants. These tests should be performed by qualified laboratory technicians.

Sensitivity Comparison Test
When performing a sensitivity comparison test, the penetrant is applied to one-half of the reference block, and the reference or control penetrant is applied to the remaining half. The processing used, including dwell time, emulsification or removal, and developing, is recommended by the penetrant manufacturer.

The indications are then visually compared under the appropriate lighting (normal or white light for visible dye penetrant indications and UV radiation for fluorescent indications). If a noticeable difference exists in the sensitivity or intensity of indications (as determined by visual observation), the penetrant is discarded. Likewise, if the penetrant shows evidence of contamination from dirt, it is discarded.

For the visual testing (VT) of fluorescent indications, the UV radiation source must have an intensity of at least 1000 mW/cm^2 at the test surface.

Water Washability Test

To achieve maximum contrast between indications and background, excess surface penetrant must be readily removable. In the water washability test, the performance of the penetrant is compared to that of the reference penetrant.

The penetrants are applied to separate test panels. After normal dwell and draining periods and emulsification (if applicable), the penetrants are washed from the panels using a uniform water spray. If washing is found to be difficult, or retention of background penetrant is noticeably different from that of the reference penetrant, the penetrant is discarded.

Water Content Test

Water content of a penetrant is best determined by the test described in *ASTM D95-13e1: Standard Test Method for Water in Petroleum Products and Bituminous Materials by Distillation*. Approximately 3.4 oz (100 mL) of the penetrant is placed in a boiling flask with a similar quantity of moisture-free xylene. The flask is connected to a reflux condenser so that the condensate drops into a 0.8 oz (25 mL) graduated tube where the water settles out. When no more water is gathered in the graduated tube (usually after a period of 1 h), the boiling process is terminated. After cooling, the volume of water in the graduated tube is read. The volume in milliliters is the percent of water by volume present in the penetrant.

The penetrant is also examined for evidence of gelling, separating, clouding, coagulating, or floating of water on the penetrant surface. If any of these conditions exist, or the percent of water exceeds specification requirements, the penetrant is discarded.

Fluorescent Luminance Test

The fluorescent luminance test is a comparison of the luminance of the control penetrant sample to the used penetrant in the tanks. If a visual comparison is made, differences of 25% are obvious, differences of 12% are noticeable and differences of 6% are detectable by the eye. Experts may sometimes detect 3% differences, but these are not typically detectable to the average observer. Most specifications require the luminance test be performed with a fluorometer (Figure 4).

In this test, a small amount of the penetrant to be tested and the reference

Figure 4. A laboratory fluorometer, specified by AMS 2644, ASTM E1417, and ISO 3452-2, measures penetrant brightness.

penetrant are diluted with a nonfluorescent, highly volatile solvent such as acetone. Test papers, cut to fit the sample holder of the fluorometer, are then dipped into the solutions, withdrawn, and dried. The samples are then alternately read on the fluorometer and the results compared. If the fluorescent intensity of the penetrant should drop below 90% of the reference penetrant, the penetrant is discarded.

Emulsifier Tests

Emulsifiers are usually tested for their sensitivity, washability, water content, and the amount of contamination from penetrants. As a general rule, the tests should be performed by laboratory technicians in accordance with ASTM specifications.

The sensitivity test, water washability test, and water content test for emulsifiers are identical to the tests used for penetrants.

Developer Tests

Developer is checked periodically for evidence of cracking, contamination, and concentration.

Dry Developer

Dry developers used in open containers or chambers are usually tested by observation. Because they are not hygroscopic, they do not absorb moisture from the air and they are relatively trouble-free if they do not come in contact with water. Any dry developer that is found lumpy or caked instead of light and fluffy, or that shows any other sign of having been wet, is discarded.

Dry developer is checked daily for caking. Additionally, the developer is visually inspected for dirt. It is also checked under UV radiation for fluorescent penetrant contamination or dirt by spreading a thin layer of developer in a 10 cm (4 in.) diameter circle. Should either condition exist, the developer is discarded.

Wet Developer

Wet (aqueous) developers are usually tested for proper concentration and possible contamination from dirt or penetrant. Solution concentration is tested by measuring the specific gravity with a hydrometer. If the hydrometer reading differs from manufacturers' requirements, either powder or water is added to the developer in sufficient quantities to bring the concentration within acceptable limits. Additionally, a small smooth test plate with no indications is dipped in the developer tank and visually examined for dirt.

The tank and the plate are checked for fluorescence or penetrant contamination under UV radiation. If either condition is in evidence, the developer is discarded.

8

Test Procedures and Standards

Procedures, Standards, and Codes

There are several different standards for PT in industrial construction, manufacturing, and maintenance applications. There are minor differences in these standards that are easily incorporated into a single procedure that covers the two types and four methods of PT.

A work instruction is task-specific and describes how to perform the process for a particular application and/or technique. A procedure is a written instruction that outlines the requirements for each step of the process. The procedure is written to comply with an industry standard or code. A standard is a document to control and govern practices in an industry or application, applied on a national or international basis, and usually produced by consensus. This relationship is outlined in Figure 1.

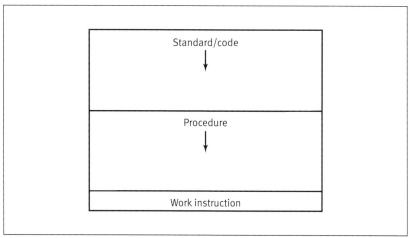

Figure 1. Relationship of standards/codes to procedures and work instructions.

ASTM E1417

ASTM E1417, Standard Practice for Liquid Penetrant Testing, provides minimum requirements for controlling the application of the liquid penetrant method. *ASTM E1417* does not provide specific instructions for conducting the inspection and, therefore, requires inspection companies to establish a penetrant inspection procedure that conforms to the standard. The company procedure requires a Level III approval.

ASTM E165
ASTM E165, Standard Practice for Liquid Penetrant Examination for General Industry, covers procedures for penetrant examination of nonporous, metallic materials, ferrous and nonferrous metals, and nonmetallic materials such as nonporous ceramics, as well as certain nonporous plastics and glass. *ASTM E165* is referenced within *ASTM E1417* for the development of detailed requirements.

ASME Boiler and Pressure Vessel Code, Section V, Article 6
The *ASME Boiler and Pressure Vessel Code* is an international code that meets the criteria for and is an American National Standard. Section V outlines the requirements for nondestructive examination. This section is divided further into articles, each of which outlines both general requirements and specific requirements for various nondestructive examination methods.

Acceptance Criteria
Examples of typical liquid penetrant acceptance criteria are as follows (sizes will vary depending on test object application).
- No linear indications allowed. A linear indication is three or more times as long as it is wide. Some specifications specify that all indications less than 0.06 in. (0.15 cm) be evaluated as rounded indications. Some specifications specify that all indications 0.02 in. (0.05 cm) and smaller be disregarded.
- No rounded indications greater than 0.03 in. (0.08 cm) for materials up to 0.25 in. (0.6 cm) design material thickness, and no rounded indications greater than 0.09 in. (0.23 cm) for materials greater than 0.25 to 0.5 in. (0.6 to 1.3 cm) design material thickness. No rounded indications greater than 0.3 in. (0.1 cm) for materials with a design material thickness greater than 0.5 in. (1.3 cm).
- The total area of acceptable rounded indications should not exceed 0.375% of the total area of the test object or weld in any 6 in.^{2} (255 cm^2) area, and there should be no more than five indications in any 1 in.^{2} (6.25 cm^2) area.
- Indications separated by less than the length or diameter of the largest indication should be considered a single indication measured from the extremities of the two indications.

Basic Methods of Instruction
Each facility should evaluate the methods of instruction and classes that are required for its PT trainees, Level I, Level II, and Level III technicians, and other personnel involved with the penetrant process. Training methods may include formal classroom lectures, video presentations, informal demonstrations, and hands-on processing of test objects.

Experience has shown that repetition of key points, hearing and reading the same concepts, and hands-on practice are valuable training tools. All training for eventual certification should be documented in the company written practice as described in *SNT-TC-1A*, with a formal course outline and the hours documented for each trainee. A method for documentation of work experience hours must also be completed before certification to Level I, Level II, or Level III.

Conclusion

NDT students completing this classroom training book will have taken an important step, from unaided VT to that of an NDT method that is capable of developing discontinuities invisible to the unaided eye.

Additionally, this classroom training book provides background critical practices for control and qualification of applied liquid penetrant technology.

Students may add technical knowledge in other NDT technologies to this basic understanding of process qualifications and controls. The results are the uniform and repeatable tests that are a hallmark of an individual certified as a Level II in PT.

References

The following publications are training references for further study and as recommended in *ANSI/ASNT CP-105: ASNT Standard Topical Outlines for Qualification of Nondestructive Testing Personnel* (2016).

AIA, 2014, *NAS 410, NAS Certification and Qualification of Nondestructive Test Personnel*, Aerospace Industries Association, Arlington, VA.

ASME, 2017, *ASME Boiler and Pressure Vessel Code*, American Society of Mechanical Engineers, New York, NY.

ASNT, 2016, *ANSI/ASNT CP-189: Standard for Qualification and Certification of Nondestructive Testing Personnel*, American Society for Nondestructive Testing, Columbus, OH.

ASNT, 1999, *Nondestructive Testing Handbook*, third edition: Vol. 2, *Liquid Penetrant Testing*, American Society for Nondestructive Testing, Columbus, OH.

ASNT, 2016, *Nondestructive Testing Handbook*, fourth edition: Vol. 1: *Liquid Penetrant Testing*, American Society for Nondestructive Testing, Columbus, OH.

ASNT, 2016, *Recommended Practice No. SNT-TC-1A: Personnel Qualification and Certification in Nondestructive Testing*, American Society for Nondestructive Testing, Columbus, OH.

ASTM, 1999, *ASTM D95-13e1: Standard Test Method for Water in Petroleum Products and Bituminous Materials by Distillation*, ASTM International, West Conshohocken, PA.

ASTM, 2010, *ASTM E1209: Standard Practice for Fluorescent Liquid Penetrant Testing Using the Water-Washable Process*, ASTM International, West Conshohocken, PA.

ASTM, 2016, *ASTM E1417: Standard Practice for Liquid Penetrant Testing*, ASTM International, West Conshohocken, PA.

ASTM, 2012, *ASTM E165: Standard Practice for Liquid Penetrant Examination for General Industry*, ASTM International, West Conshohocken, PA.

General Dynamics, 1981, *Nondestructive Testing Classroom Training Handbook*, fourth edition: *Liquid Penetrant Testing*, Convair Division of General Dynamics Corp., Fort Worth, TX.

Henry, E., 2018, *Programmed Instruction Series: Liquid Penetrant Testing*, American Society for Nondestructive Testing, Columbus, OH.

ISO, 2013, *ISO 3452-2: Non-destructive Testing – Penetrant Testing – Part 2: Testing of Penetrant Materials*, Geneva, Switzerland.

SAE, 1996, *SAE AMS 2644: Inspection Material, Penetrant*, SAE International, Warrendale, PA.

Figure Sources

All figures derive from sources published by The American Society for Nondestructive Testing Inc., except for the following:

Chapter 4
Figure 8 – Shutterstock / Phakdee Kasamsawad
Figures 9 and 10 – Magnaflux
Figures 13, 14, and 15 – Peter Pelayo, Met-L-Chek

Chapter 5
Figure 2 – NDI branch, Materials Division, Naval Air Warfare Center Aircraft Division at the Patuxent River Naval Air Station, 2016

Chapter 6
Figures 5, 9, 10, and 12 – Peter Pelayo, Met-L-Chek

Glossary

Aqueous developer: See *Wet developer*.

Background: The surface on which an indication is viewed. It may be the natural surface of the test object, or it may be the developer coating on the surface. This background may contain traces of unremoved penetrant (fluorescent or visible), which if present, can interfere with the visibility of indications.

Background fluorescence: Fluorescent residue observed over the general surface of the test object during a fluorescent penetrant test.

Bath: Term used colloquially to designate the liquid penetrant testing materials into which test objects are immersed during the testing process.

Black light: See *Ultraviolet radiation*.

Bleedout: The action of the entrapped penetrant in spreading out from surface discontinuities to form an indication.

Blotting: The action of the developer in soaking up penetrant from a surface discontinuity, so as to cause maximum bleedout of penetrant for increased contrast and sensitivity.

Capillary action: The tendency of liquids to penetrate or migrate into small openings, such as cracks, pits or fissures.

Clean: Free of interfering solid or liquid contamination.

Color contrast penetrant: See *Visible dye penetrant*.

Comparative reference block: An intentionally cracked metal block having two separate but adjacent areas for the application of different penetrants so that a direct comparison of their relative effectiveness can be obtained. Can also be used to evaluate penetrant test techniques and test conditions.

Contact emulsifier: A fluid that begins emulsifying penetrant on simple contact with the penetrant. Usually oil based (lipophilic).

Contrast: The difference in visibility (illumination or coloration) between an indication and the surrounding surface.

Dark adaptation: The adjustment of the eyes when one passes from a bright to a darkened area.

Defect: A discontinuity that interferes with the usefulness of an object. A fault in a material or test object that is detrimental to its serviceability.

Detergent remover: A penetrant remover that is a solution of a detergent in water. Also see *Hydrophilic emulsifier*.

Developer: A material that is applied to the test object surface after excess penetrant has been removed and that is designed to enhance the penetrant bleedout to form indications. The developer may be a fine powder, a solution that dries to a fine powder or a suspension (in solvent, water, alcohol, and so on) that dries leaving an absorptive film on the test surface.

Developing time: The time between the application of the developer and the examination of the test object for indications. The elapsed time necessary for the applied developer to bring out indications from penetrant entrapments. Also called development time.

Discontinuity: An interruption in the normal physical structure or configuration of an object, such as cracks, forging laps, seams, inclusions, porosity, and so on. A discontinuity may or may not affect the usefulness of the test object.

Dragout: The loss of penetrant materials from a tank as a result of their adherence to the objects being processed.

Drain time: That portion of the penetrant testing process during which the excess penetrant, emulsifier, detergent remover, or developer is allowed to drain from the test object.

Dry developer: A fine dry powder developer that does not use a carrier fluid.

Drying oven: An oven used for drying rinse water from test objects.

Drying time: The time allotted for a rinsed test object to dry.

Dual sensitivity penetrant: A penetrant that contains a combination of visible and fluorescent dyes.

Dwell time: The total time that the penetrant or emulsifier is in contact with the test surface, including the time required for application and the drain time. Also see *Emulsification time*.

Electrostatic spraying: A technique of spraying wherein the material being sprayed is given a high electrical charge, while the test object is grounded.

Emulsification time: The period of time that an emulsifier is permitted to combine with the penetrant before removal. Also called emulsifier dwell time.

Emulsifier: A liquid that combines with an oily penetrant to make the penetrant water-washable. Also see *Hydrophilic emulsifier* and *Lipophilic emulsifier*.

Evaluation: The process of determining the severity of the condition after the indication has been interpreted. Evaluation leads to determining whether the object is acceptable, salvageable or rejectable.

False indication: An indication caused by improper processing. Distinct from nonrelevant indication.

Flash point: The lowest temperature at which a volatile flammable liquid will give off enough vapor to make a combustible explosive mixture in the air space surrounding the liquid surface.

Flaw: See *Discontinuity*.

Fluorescence: The emission of visible radiation by a substance as the result of, and only during, the absorption of ultraviolet radiation.

Fluorescent penetrant: A testing penetrant that is characterized by its ability to fluoresce when excited by ultraviolet radiation.

Hydrophilic emulsifier: A water-based agent that, when applied to an oily penetrant, renders the penetrant water-washable. Can be used as a contact emulsifier, but more often the emulsifier is added to the water rinse and accompanied by some form of mechanical agitation or scrubbing to remove excess penetrant. Sometimes called a hydrophilic remover.

Indication: That which marks the presence of a discontinuity, as the result of detectable bleedout of penetrant from the discontinuity.

Inspection: The visual examination of a test object after completion of the penetrant processing steps.

Interpretation: The determination of the significance of indications from the standpoint of whether they are relevant or nonrelevant.

Leak testing: A technique of liquid penetrant testing in which the penetrant is applied to one side of the surface, while the other side is tested for indications that would indicate a leak or void.

Lipophilic emulsifier: An oil-based agent that, when applied to an oily penetrant, renders the penetrant water-washable. Usually applied as a contact emulsifier.

Nonaqueous wet developer: A developer in which the developing powder is applied as a suspension in a quick drying solvent. Also called solvent developer.

Nonfluorescent penetrant: See *Visible dye penetrant*.

Nonrelevant indication: An indication that is not or cannot be associated with a discontinuity.

Penetrability: The property of a penetrant that causes it to find its way into very fine openings, such as cracks.

Penetrant: A liquid capable of entering discontinuities open to the surface, and which is adapted to the testing process by being made highly visible in small traces. Fluorescent penetrants fluoresce brightly under ultraviolet radiation, while the visible penetrants are intensely colored to be noticeable under visible light.

Penetration time: See *Dwell time*.

Postemulsification: A penetrant removal technique using a separate emulsifier.

Postemulsifiable penetrant: A penetrant that requires the application of a separate emulsifier to render the surface penetrant water-washable. Can be removed by applying a solvent remover.

Precleaning: The removal of surface contaminants or smeared metal from the test object so that they cannot interfere with the penetrant testing process.

Quenching of fluorescence: The extinction of fluorescence by causes other than removal of ultraviolet radiation (the exciting radiation).

Rinse: The process of removing liquid penetrant testing materials from the surface of a test object by washing or flooding with another liquid—usually water. Also called wash.

Self-emulsifiable: See *Water-washable penetrant*.

Sensitivity: The ability of the penetrant process to detect surface discontinuities.

Solvent developer: See *Nonaqueous wet developer*.

Solvent removed: A penetrant removal technique wherein the excess penetrant is wiped from the test surface with a solvent remover.

Solvent remover: A volatile liquid used to remove excess surface penetrant from the test object. Sometimes called penetrant remover.

Surface tension: That property of liquids which, because of molecular forces, tends to bring the contained volume into a form having the least superficial area.

System: With respect to liquid penetrant testing materials, a combination of liquid penetrant and emulsifier that are furnished by the same manufacturer and are qualified together. For water-washable and solvent-removable liquid penetrants, a system consists of the liquid penetrant only.

Ultraviolet radiation (for liquid penetrant testing): Light radiation in the near ultraviolet range (UV-A) of wavelengths (320 to 400 nm), just shorter than visible light.

Ultraviolet radiation filter: A filter that transmits near ultraviolet radiation while suppressing visible light and harmful ultraviolet radiation.

Visibility: The characteristic of an indication that enables the observer to see it against the conditions of background, outside light, and so on.

Viscosity: The state or degree of being viscous. The resistance of a fluid to the motion of its particles.

Visible liquid penetrant: A testing penetrant that is characterized by its intense visible color – usually red. Also called color contrast or nonfluorescent penetrant.

Wash: See *Rinse*.

Water-soluble developer: A developer in which the developer powder is dissolved in a water carrier to form a solution. Not a suspension.

Water-suspendable developer: A developer in which the developer particles are mixed with water to form a suspension.

Water wash: A penetrant removal technique wherein excess penetrant is washed or flushed from the test surface with water.

Water-washable penetrant: A type of penetrant that contains its own emulsifier, making it water-washable.

Water tolerance: The amount of water that a penetrant, emulsifier can absorb before its effectiveness is impaired.

Wet developer: A developer in which the developer powder is applied as a suspension or solution in a liquid – usually water or alcohol.

Wettability: The ability of a liquid to spread out spontaneously and adhere to solid surfaces.

Index

Page numbers for photographs or illustrations are in *italics*. Page numbers for tables or charts are in **bold**. Liquid penetrant testing may be abbreviated PT, and the American Society for Nondestructive Testing may be abbreviated ASNT.

A
acceptance criteria, 20–21, 82
acetone, 14–15
aerosol spray cans for portable methods, 26
aerospace components
    fluorescent PT light requirements, 42–43
    Method B prohibited, 51
    Type I systems for, 26
    V-22 Naval aircraft part using Type I Method D, 51, *52*
agitation for water-suspendable developers, 19
air pollution, 9
alkaline detergent for steam cleaning, 13
aluminum reference blocks for quality control tests, 75–76, *75*
anodized test panels for quality control tests, 76–77, *77*
applicators and sprayers, 36
approval certificates, 11
ASME Boiler and Pressure Vessel Code, 82
ASTM D95-13e1: Standard Test Method for Water in Petroleum Products and Bituminous Materials by Distillation, 79
ASTM E165, Standard Practice for Liquid Penetrant Examination for General Industry, 25, 82
ASTM E1417, Standard Practice for Liquid Penetrant Testing, 25, 81
ASTM International, 25, 73
automatic processing, dry powder developers for, 18, 45
auxiliary equipment, 36–38, *37–38*

B
basic principles of PT, 4–5, *5*
blast, 14
blowholes in plate, 67
bursts in forging, 63–64

C
capillary action, 59
castings, 63–66, *64–66*
    dry powder developers for, 45
ceramic reference blocks for quality control tests, 76
cerium-activated calcium phosphate phosphor, 42
certification, 6–8, 60, 83
classification system for PT, **6**
cold shuts, 23, 66, *66*
cold-working of surfaces, 14
color contrast penetrants, 44
commercially available PT materials, 6
contact emulsifiers, 44
contamination
    in bulk materials, 43
    cross-contamination in reference blocks and panels, 75–76
    false indications from, 69–70
    organic contaminants, 13
    in PT materials, 78, 80
    quality control procedures for, 73
    risk reduction in aerosol cans, 26, 53
    sources, 21, *21*, 69–70
continuous linear indications, 22–23, *23*
control sample storage and handling, 73
cracks, defined and described, 23, 62
crater cracks/crater pits, 67, *68–69*

D
darkened inspection booths, 42
deep cracks, 23–24
detergent cleaning, 12–13
developers
    advantages and disadvantages, 53–54
    application and drying, 17–19
    development time, 20
    quality control tests, 80
die marks, 23
dip tank applications for emulsification, 27, 28, 31
discontinuities
    categories, 55–58, *55–59*
    depth determination, 24
    evaluation of, 20–21
    fine discontinuities, 45
    indication formation, 58–59
    inherent discontinuities, 55–56, *55–56*
    internal discontinuities, 64, 66
    interpretation of, 21
    LCF block standards, 78
    metal-smearing effects on, 14, 60, *60*
    permanent records of, 54
    principles for detection, 4–5, *5*

processing discontinuities, 56, *56–57*
service discontinuities, 58, *58–59*
shallow discontinuities, 27, 51
as stress risers, 23
surface discontinuities, 3, 11
See also indications
dry powder developers, 17–18, *18*, 45, 80
dryer ovens, 17
drying test objects, 14–15
dust cloud chambers, 18
dwell time, defined and described, 15

E

ecological requirements for material disposal, 46
electromagnetic spectrum, *40*
electrostatic sprayers, 18
emery cloth, 14
employee certification, 6–8, 60, 83
emulsifiers and emulsification
    emulsifier tests for quality control, 80
    lipophilic emulsification method, 28, *29*
    UV radiation background lighting for, 15
    for water-washability, 44
environmental concerns, 13
equipment. See liquid penetrant testing units
etching for metal smearing, 11, 14, *14*, 23, 60
evaluation, 20–22, 59
examinations for technician levels, 8
excess surface penetrant removal, 15–20
exhaust fans, 36
experience for employee certification, 8
extrusions, 69

F

fading of penetrant dyes, 78
faint indications, *23*, 24
false indications, 20, 21, *21*, 69–70, *70*
field tests with portable penetrant kits, 26
fine discontinuities, 45
fine threads, 45
fire concerns, 8–9
flammable liquid precautions, 46
fluorescent luminance test for quality control, 79–80, *79*
fluorescent or visible methods, 25–26
fluorescent penetrants
    false indications with, 21
    fluorescent penetrant kit, 39, *39*
    illumination requirements, 15
    UV radiation source for, 15
    visibility of, 44
    visible dye penetrants and, 61
    washability testing, 77–78
fluorescent UV radiation tubes, 41
fluorometers, 79–80, *79*
food-compatible materials, 45
footcandles (lux), 43
forging laps, 23, 63–64, *63*
Form A dry developer, 53

Form B water-soluble, 53
Form C water-suspendable, 53
Forms D (fluorescent) and E (visible) nonaqueous, 53–54
Form F special application, 54
fusion line cracks in welds, 69

G

grit blasting, 11
gross indications, *23*, 24

H

halogen-free cleaners, 44
health hazard precautions, 46
high temperature penetrants, 45
higher sensitivity penetrants (3, 4), 53
high-temperature materials, 59–60
history of PT, 3, *3*
hot tears in castings, 66, *66*
hydro-air nozzles, 16
hydrometers, 36, *38*, 80
hydrophilic (water-based) emulsification method, 30–31, *30*, 44

I

illumination requirements, 15–16, 43, 60
immersion tanks, 12, *12*, 14, *14*
immersion testing, 18–19
inclusions in plate, 67
indications, 55–72
    castings, 64–66, *64–65*
    crack indications, 62
    discontinuity categories, 55–58
    discontinuity indication formation, 58–59
    evaluation of, 69–72
    extrusions, 69
    factors affecting, 60
    false indications, 69–70, *70*
    forging laps, 63–64, *63*
    inherent discontinuities, 55–56, *55–56*
    interpretation, 21–24, 58–69
    lighting, 60
    metal smearing, 60, *60*
    nonrelevant indications, 71–72, *71–72*
    penetrant use in indications, 60–61, *61*
    persistence of, 59
    plate, 66–67, *67*
    porosity indications, 63
    prior processing, 61
    processing discontinuities, 56–57, *56–57*, 62
    relevant indications, 70–71, *71*
    sequence, 60
    service discontinuities, 57, *57–59*, 62, *62–63*
    solidification cracks, 62
    specific material form indications, 63
    surface condition, 61–62
    temperature effects, 59–60
    test object preparation, 60
    time for indication appearance, 59

types, 20–21
welds, 67–69, *67–68*
See also discontinuities
ingots, discontinuities and indications in, 55, 62, 63
inherent discontinuities, 55–56, *55–56*
inhibited solvent removers, 45
inline penetrant systems, 26–27, *26*
inspection booths, 42
instruction methods, 82–83
intermittent or broken line indications, 23, *23*
internal discontinuities, 66
interpretation, 20
isopropyl alcohol, 14–15

L

laminations in plate, 66–67, *67*
lamps, 36
laps in forging, 63, *63*
LED (light-emitting diode) lamps, 41–42, *41*
Level I and II technicians, techniques and procedures, 11
Level II technicians, evaluations, 21
Level III technicians, approval certificates, 11
light meters, 43, *43*
light-emitting diode (LED) lamps, 41–42, *41*
lighting standards, 15–16, 43, 60
lipophilic (oil-based) emulsification method, 27–28, *27*, *28*, 44
liquid dip tanks, 18–19
liquid honing, 14
liquid oxygen environments, 24
liquid penetrant testing (PT)
    air pollution, 9
    basic principles, 4–5, *5*
    certification, 8
    classification system, **6**
    commercially available PT materials, 6
    employee certification, 7–8
    examination, 8
    experience, 8
    fire concerns, 8–9
    history, 3, *3*
    personnel qualification and certification, 6–7
    purpose, 4
    safety precautions, 8–9
    skin irritation, 9
    training, 8
    ultraviolet radiation (UV), 9
liquid penetrant testing units, 33–46
    auxiliary equipment, 36–38, *37–38*
    dry developers, 45
    emulsifiers, 44
    exhaust fans, 36
    fluorescent penetrant kit, 39, *39*
    hydrometers, 36, *38*
    lamps, 36
    light meters, 43, *43*
    materials for, 43–46
    nonaqueous wet developer, 45
    portable equipment, 38–39

    postemulsification penetrants, 44
    precautions, 46
    precleaning and post-cleaning materials, 44
    pumps, 36
    refractometers, 36, *38*
    safety concerns for UV lamps, 42
    solvent removers, 45
    solvent-removable penetrants, 44
    special purpose penetrant materials, 45–46
    sprayers and applicators, 36
    stations, 34–35, *34–35*
    subdued white light, testing in, 42–43
    thermostats and thermometers, 36
    timers, 36
    tubular fluorescent cold discharge sources, 42
    ultraviolet radiation, 39–43, *40–41*
    visible dye penetrant test kit, 38, *39*
    water-based developers, 45
    water-washable penetrants, 44
low chlorine materials, 45
low cycle fatigue (LCF) blocks, 78
low energy emulsifiers, 45
low sulfur materials, 45
lower sensitivity penetrants (1/2, 1), 52

M

magnetic particle testing (MT), 64
magnification for small indications, 61, *61*
maintenance shops, inline penetrant systems for, 26
manual wipe (Method C), 16–17, *17*, 26, 29
manufacturers' recommendations
    compliance with, 49, 53
    LCF block cleaning, 78
    quality control of test materials, 73
    safety procedures and, 45
manufacturing process, indications resulting from, 63
material discontinuities. *See* discontinuities
material tests for quality control, 78–80
materials, 43–46
    dry developers, 45
    emulsifiers, 44
    nonaqueous wet developers, 45
    postemulsification penetrants, 44
    precleaning and post-cleaning materials, 44
    solvent removers, 45
    solvent-removable penetrants, 44
    special purpose penetrant materials, 45–46
    water-based developers, 45
    water-washable penetrants, 44
mechanical cleaning, metal smearing with, 60, *60*
metal scrapers, 14
metal smearing, 11, 14, 23, 60, *60*
metals, inherent discontinuities in, 55–56, *55–56*
Method A. *See* water-washable method
Method B. *See* lipophilic (oil-based) emulsification method
Method C. *See* manual wipe method; solvent-removable method
Method D. *See* hydrophilic (water-based) emulsification method

method characteristics, 25–26
method selection, 49–54
    advantages and disadvantages, 50–54, **50**
    developers, 53–54
    factors in, 49, *49*
    Form A dry developer, 53
    Form B water-soluble, 53
    Form C water-suspendable, 53
    Forms D (fluorescent) and E (visible) nonaqueous, 53–54
    Form F special application, 53–54
    higher sensitivity penetrants (3, 4), 53
    lower sensitivity penetrants (1/2, 1), 52
    Method A water-washable method, 51
    Method B postemulsifiable (lipophilic) method, 51
    Method C solvent-removable method, 51
    Method D postemulsifiable (hydrophilic) method, 51, *52*
    penetrant method advantages and disadvantages, 51
    penetrant sensitivity, 52, *52*
    penetrant types, 50
    responsibility for, 49
    Type I fluorescent penetrant, 50
    Type II visible penetrant, 50
    Type III dual-mode (visible and fluorescent) penetrant, 50
microscope examination, 61
microwatts per centimeter squared ($\mu W/cm^2$), 43, *43*
moisture in discontinuity cavity, 61–62

N

nickel-chromium-iron (NiCrFe) plates, 78
nonaqueous wet developer, 19, 45
nonrelevant indications, 20, 22, 71–72, *71–72*

O

oil-based emulsification method. *See* lipophilic (oil-based) emulsification method
organic contaminants, 12, 13
over rinsing, 44
overhaul shops, inline penetrant systems for, 26

P

paint removal, 13, *13*
peening, 14
penetrants
    application techniques, 15
    dyes, fading of, 78
    penetrant use in indications, 60–61, *61*
    removal of excess, 15–20
    sensitivity, 52, *52*
    types, method selection for, 50–53
    See also liquid penetrant testing units; *specific penetrant types*
persistence of indications, 59
personnel qualification and certification, 6–7
plastic film developers, 45
plate indications, interpretation of, 66–67, *67*
plated test panels for quality control tests, 76–77, *77*
porosity, 23–24, 63, 67, *67–68*
portable equipment, 38–39
post-cleaning, 12–15, 24
postemulsification penetrants, 44
postemulsified method, 61
power wire brushing, 11, 14, 60, *60*
precautions with materials, 46
precleaning and post-cleaning equipment and materials, 12–15, 44
precleaning of test objects, 11–12, 14
preparation of test objects, 11
prior processing indications, 61
procedures, defined and described, 11, 81, *81*
procedures, standards, and codes, 81–82
procedures and techniques, 11–12
process control, 73–80
    aluminum reference blocks, 75–76, *75*
    anodized test panels, 76–77, *77*
    ceramic reference blocks, 76
    control samples, 73
    developer tests, 80
    dry developer tests, 80
    emulsifier tests, 80
    fluorescent luminance test, 79–80, *79*
    liquid penetrant material tests, 78–80
    low cycle fatigue blocks, 78
    plated test panels, 76–77, *77*
    quality control of test materials, 73
    reference blocks, 73–78
    sensitivity comparison test, 78
    stainless steel panels, 77–78
    system monitor panels, 74–75, *74*
    test material control samples, 73
    twin nickel-chromium sensitivity panels, 77
    water content test, 79
    water washability test, 79
    wet developer tests, 80
processing, 11–24
    continuous linear indications, 22–23, *23*
    detergent cleaning, 12–13
    developer application and drying, 17–19
    development time, 20
    discontinuity depth determination, 24
    dry powder developers, 17–18, *18*
    drying test objects, 14–15
    dwell time, 15
    etching, 14, *14*
    evaluation, 20–21
    excess surface penetrant removal, 15–20
    faint indications, *23*, 24
    false indications, 21, *21*
    gross indications, *23*, 24
    illumination, 15
    indications, 21–24
    intermittent or broken line indications, 23, *23*
    interpretation, 20
    manual wipe (Method C), 16–17, *17*

nonaqueous developers, 19
nonrelevant indications, 22
paint removal, 13, *13*
penetrant application, 15
post-cleaning, 24
precleaning and post-cleaning equipment, 12–15
precleaning of test objects, 11–12
precleaning processes to avoid, 14
preparation of test objects, 11
procedures and techniques, 11–12
relevant indication categories, 22–24
relevant indications, 22
rounded indications, 23–24, *23*
rust and surface scale removal, 13
solvent cleaning, 12, *12*
steam cleaning, 13
ultrasonic cleaning, 13
vapor degreasing, 13
water rinse methods, 16, *16*
water soluble developers, 18–19
water-suspendable developers, 19
processing cracks, 62
processing discontinuities, 56–57, *56–57*
prohibited cleaning methods, 11
pumps, 36
purpose of PT, 4

Q

quality control. *See* process control

R

radiographic testing (RT), 64, 66
reference blocks for quality control tests, 73–78
refractometers, 31, 36, *38*
relevant indications, 20, 22–24, 70–71, *71*
rough surfaces, 18, 22, 45, 53, 61
rounded indications, 23–24, *23*
RT (radiographic testing), 64, 66
rust and surface scale removal, 13

S

SAE AMS 2644, Inspection Material, Penetrant, 25
SAE International, 25
safety concerns and precautions, 8–9, 13, 42
sand blasting, metal smearing with, 60, *60*
scratches, 23
seams, 23
sensitivity
    dry powder developers, 18
    emulsifier tests, 80
    of Form A dry developer, 53
    of Form D fluorescent developer, 53
    method comparisons, 6, **6**
    nonaqueous developers, 19
    penetrant choice for, 25–26, 60–61
    sensitivity comparison test for quality control, 78
sequence, 60
service cracks, 62, *62–63*
service discontinuities, 57, *57–59*

shotpeening, 11
shrinkage in castings, 64, *65*
skin irritation, 9
slag in plate, 67
SNT-TC-1A, 83
solidification cracks, 62
solvent cleaning, 12, *12*
solvent- or water-removable method for field tests, 26
solvent removers, 45
solvent-removable method (Method C), 29, *29*
solvent-removable penetrants, 21, 44
special purpose penetrant materials, 45–46
specific gravity of liquid, 18, 19
specific material form indications, 63
sprayers and applicators, 36
stainless steel panels for quality control tests, 77–78
standard methods, 25–31
    fluorescent or visible methods, 25–26
    hydrophilic (water-based) emulsification method (Method D), 30–31, *30*
    inline penetrant systems, 26–27, *26*
    lipophilic (oil-based) emulsification method (Method B), 27–28, *27, 28*
    method characteristics, 25–26
    solvent- or water-removable method for field tests, 26
    solvent-removable method (Method C), 29, *29*
    water-washable method (Method A), 27–28, *27*
standards
    defined and described, 81, *81*
    See also test procedures and standards
stations, 34–35, *34–35*
steam cleaning, 13
stress risers, 23
subdued white light, testing in, 42–43
sulfur-free cleaners, 44
supersensitive penetrants, 45
surface condition, 61–62
surface scale removal, 13
system monitor panels for quality control tests, 74–75, *74*

T

techniques, defined and described, 11
temperature effects, 59–60
test material control samples, 73
test objects
    dished or hollowed areas, 19
    precautions, 46
    precleaning, 11–12
    preparation, 11, 60
    properties and condition, 11
test procedures and standards, 81–83
    acceptance criteria examples, 82
    instruction methods, 82–83
    procedures, standards, and codes, 81–82
testable surfaces, 11
testing units. *See* liquid penetrant testing units
thermostats and thermometers, 36
time for indication appearance, 59

timers, 36
titanium plates for low cycle fatigue block cracks, 78
training, 8
transverse cracks in welds, 69
tubular fluorescent cold discharge sources, 42
2024-T3 aluminum alloy plate, 75–76
twin nickel-chromium sensitivity panels for quality control tests, 77
Type I fluorescent penetrant, 16, 50
Type I systems emulsification time, 28
Type II systems emulsification time, 28
Type II visible penetrant, 16, 50
Type III dual-mode (visible and fluorescent) penetrant, 50

U

ultrasonic cleaning, 13
ultrasonic testing (UT), 24, 61, 64, 67
ultraviolet radiation (UV)
    background lighting for fluorescent penetrant application, 15
    for fluorescent liquid penetrant indications, 39–43, *40–41*
    safety concerns and precautions, 9
    UV mercury vapor arc lamps, 39–41, *40*
UT (ultrasonic testing), 24, 61, 64, 67

V

V-22 Naval aircraft part using Type I Method D, 51, *52*
vapor degreasing, 13
visible dye penetrants
    fluorescent penetrant application and, 61
    illumination requirements, 15
    test kit, 38, *39*
    washability testing, 77–78
    water-washable, 44
volatile liquid solvents, 19

W

washability testing (emulsifier tests), 80
water content test, 79–80
water in discontinuity cavity
    blockage of penetrant, 61–62
    evaporation, 17
    water-soluble solvent for, 14–15
water rinses, 16, *16*, 26, 44
water soluble developers (liquid dip tanks), 18–19
water washability test for quality control, 79
water-based developers, 45
water-removable method, 26
water-soluble developers, 45
water-suspendable developers, 19, 45
water-washable method (Method A), 27–28, *27*
water-washable penetrants, 44
wax film developers, 45
welds
    fusion lines, 69
    high temperature penetrants for, 45
    indication types, 63, 67–69, *67–68*
    root layer of, 60
    undercut, 69
wet developer tests for quality control, 80
white clothing reflection, 15
white light sensor, 43, *43*
work instruction, 81, *81*